AF762024

Climate Change and Its Causes, Effects and Prediction

nova science publishers

www.novapublishers.com

Climate Change and Its Causes, Effects and Prediction

Transforming Futures – Navigating Society, Environment, and Governance
Medani P. Bhandari, PhD (Editor)
2023. ISBN: 979-8-89113-319-8 (Hardcover)
2023. ISBN: 979-8-89113-421-8 (eBook)

Urban Heat Islands Reexamined
Satyaprakash, PhD (Editor)
Anne W. M. Ng, PhD (Editor)
2022. ISBN: 979-8-88697-215-3 (Hardcover)
2022. ISBN: 979-8-88697-287-0 (eBook)

Environmental Contamination and Climate Change: Effect on Plants and Remedial Strategies
Dhriti Kapoor, PhD (Editor)
Renu Bhardwaj, PhD
M.Phil (Editor)
Vandana Gautam, PhD (Editor)
2021. ISBN: 978-1-53619-667-2 (Hardcover)
2021. ISBN: 978-1-53619-732-7 (eBook)

Managing Climate Change
Rex S. Soto (Editor)
2021. ISBN: 978-1-53619-496-8 (Hardcover)
2021. ISBN: 978-1-53619-560-6978 (eBook)

More information about this series can be found at
https://novapublishers.com/product-category/series/climate-change-and-its-causes-effects-and-prediction/

Mohamad Nor Azra
Thirukanthan Chandra Segaran
Guillermo Téllez-Isaías
and Hidir Ariffin
Editors

Marine Life in Changing Climates

nova
science publishers
www.novapublishers.com

NOTICE TO THE READER

Library of Congress Cataloging-in-Publication Data

Names: Azra, Mohamad Nor, editor. | Segaran, Thirukanthan Chandra, editor. | Téllez-Isaías, Guillermo (Veterinarian), editor. | Ariffin, Hidir, editor.
Title: Marine life in changing climates / (editors) Mohamad Nor Azra, PhD, Institute of Climate Adaptation and Marine Biotechnology (ICAMB), Universiti Malaysia Terengganu (UMT), Kuala Nerus, Terengganu, Malaysia, Thirukanthan Chandra Segaran, PhD, Institute of Climate Adaptation and Marine Biotechnology (ICAMB), Universiti Malaysia Terengganu (UMT), Kuala Nerus, Terengganu, Malaysia, Guillermo Téllez, PhD, Professor, Department of Poultry Science, University of Arkansas, Fayetteville, AR, USA, Hidir Ariffin, PhD, Higher Institution Centre of Excellence (HICoE), Institute of Tropical Aquaculture and Fisheries, Universiti Malaysia Terengganu, Kuala Nerus, Terengganu, Malaysia.
Description: New York : Nova Science Publishers, [2024] | Series: Climate change and its causes, effects and prediction | Includes bibliographical references and index.
Identifiers: LCCN 2024002167 (print) | LCCN 2024002168 (ebook) | ISBN 9798891134041 (paperback) | ISBN 9798891135109 (adobe pdf)
Subjects: LCSH: Marine organisms--Effect of temperature on. | Marine organisms--Climatic factors. | Marine organisms--Effect of global warming on. | Marine ecosystem health.
Classification: LCC QH545.T4 M365 2024 (print) | LCC QH545.T4 (ebook) | DDC 578.77--dc23/eng/20240226
LC record available at https://lccn.loc.gov/2024002167
LC ebook record available at https://lccn.loc.gov/2024002168

Published by Nova Science Publishers, Inc. † New York

Contents

Preface

The marine environment and coastal ecosystems are critical links in the sustainability of earth life, such as for mollusks, crustaceans, and other fisheries species, and will play an important role in ecosystem services in the future. Meanwhile, the climate change scenario is having a severe impact on these natural resources, as well as their food supply and production. Because of the vast amount of literature available on online academic databases, scientometric methodologies may benefit from identifying research gaps and potential future fields of study. This book evaluates the climate risks to various types of marine life. The authors navigate through the narratives of ocean researchers, conservationists, and communities, each of whom shares their recent research and perspectives on the challenges of climate change. The book is also a call to action, highlighting the critical importance of understanding, mitigating, and adapting to the effects of climate change on marine environmental ecosystems. From university initiatives to global collaboration, this scientific book sheds light on the current situation of climate change on marine resources, offering a glimmer of hope for ocean sustainability. This book on marine life in a changing climate is not confined to the blue horizon. The next volume of this book not only covers the literature of this book but also delves deeper into the current research and challenges of our ocean in a constantly changing climate.

Chapter 1 - Generally, climate change has been defined as "any change in climate over time, whether due to natural variability or as a result of human activity." There are many elements and terminologies that have been used to identify or categorize "climate change" such as "global warming," or "extreme event," or "anthropogenic changes," or "greenhouse effects," or "weather pattern," or "sea level rise," or "coastal erosion," or "heat wave," or "acidification," or, "drought," or "flood," or "heavy rainfall," or "melt of the glacier," "hypoxia," or "harmful alga bloom," or "eutrophication," or "atmospheric pollution," or "permafrost," or "agriculture runoff" or "burning of fossil fuel." Meanwhile, the marine ecosystem is among one of the most impacted by the continuous changes in climate and concurrently receiving numerous stressors from natural and anthropogenic sources. However, limited studies of unequivocal theoretical framework on how marine life responds to climate change strongly compromises recent ability to make further predictions in the future. Thus, combining all of this knowledge on how marine life responds to climate change within the edited series is vital.

Chapter 2 - As global climate change intensifies, both human and natural ecosystems must adapt to its multifaceted impacts. Extensive research has already been conducted to gauge how these environments respond to and prepare for climate-induced alterations. Collating this knowledge is essential, as it not only offers a consolidated view of present findings but also highlights areas yet to be explored. To this end, the authors surveyed pertinent English-language articles from the Web of Science Core Collection (WOSCC) using keywords "climate adaptation" and "climate change adaptation." Utilizing the CiteSpace software, the authors constructed visual knowledge maps to evaluate the present state and developmental trajectory of climate change adaptation research. From the authors' search, the authors amassed 9,098 articles in the WOSCC, published between 1972 and 31-12-2021. Notably, around 53% of these were contributions from the USA, Australia, and England. Over the past five decades, this field has seen participation from 6,664 institutions, collaboration among 26,814 authors, and publications across 1,680 distinct journals. Through CiteSpace, the authors discerned 14 primary clusters, with prominent themes including spatial planning, nature-based solutions, and strategies for smallholder farmers. Recurring keywords encompassed "vulnerability," "management," "evolution," "geographic variation," "body size," and "temperature." The emphasis on restructuring human habitats via spatial planning and nature-based solutions underscores the scientific community's proactive approach to climate change adaptation. A significant portion of this work pertains to the realms of "Ecology," "Earth & Marine," and "Economics, Economic & Political Studies." The authors' findings advocate for policymakers to prioritize localized adaptation strategies and foster informed decision-making to adeptly navigate the challenges of climate change. This chapter serves as a valuable resource for scientific communities, philanthropic entities, relevant government bodies, and NGOs, guiding them in the sustainable management of the repercussions of climate change.

Chapter 3 - Climate change poses significant challenges to the sustainability of capture fisheries, a vital component of global food security and nutrition. In this scientometric study, the authors examine current trends and identify future research directions concerning the impact of climate change on capture fisheries and fish landings. The authors analyzed 3164 publications from the Web of Science Core Collection database utilizing quantitative metadata analysis and scientometric techniques. The authors' findings reveal that the United States and Japan are the largest contributors and exert the most influence in this research area. Highly cited articles indicate that climate change has the potential to significantly redistribute global fisheries catch, with up to a 70% shift, particularly affecting high latitude regions and the tropics. Prominent and influential keywords within the knowledge base encompass "climate change-management-impact" and "climate change-small scale fishery-impact." A cluster analysis of article titles identifies the top three research foci as future trajectories, food security, and small pelagic fish. Through the authors' scientometric investigation, the authors found that addressing the impacts of climate change on small-scale fisheries and maintaining food security under changing environmental conditions are among the most critical research priorities. By understanding the current research landscape and identifying emerging trends, researchers can better collaborate and address critical questions related to the effects of climate change on capture fisheries and fish landings, ultimately contributing to more sustainable and resilient fisheries management.

Chapter 4 - Shellfish serve as vital protein sources for humans, representing some of the primary aquacultured food resources globally. Species like molluscs are predominantly cultured in coastal regions, though there's a growing trend towards offshore mollusc farming. Both coastal and deep-sea habitats are increasingly affected by the repercussions of climate change. Elevated temperatures warm the ocean, subsequently impacting ecological dynamics and the organisms within these ecosystems. Therefore, the reliance of shellfish culture on aquatic ecosystems may affect this protein source production. This study reviews an emerging research frontier by identifying significant impacts of climate change on the shellfish, particularly molluscs. The Web of Science Core Collection (WOSCC) database was used as a main proxy to extract the bibliometric information and CiteSpace software was used to analyze the scientometric dataset. Data was generated from WOSCC from 1970 until December 31, 2021. A total of 28,061 articles were generated, with inferential statistics from a descriptive dataset showing collaborative networks between authors, institutions and countries. Meanwhile the citation data also indicated that keywords such as oxidative stress, growth, temperature, rat, hypoxia, and calcium are among the most used keywords. When examining the prevailing trends in leading publications on this topic, ocean acidification emerges as a primary concern, as corroborated by recent research. Recognizing the significance of this issue and the observed research gap concerning the effects of climate change on molluscs, this study endeavors to bridge this void. To the best of the author's understanding, this is the inaugural paper employing scientometric analysis in the context of mollusc-related research on climate change. Data retrieved from the WOSCC database indicates a consistent trajectory of publications, highlighting the burgeoning potential of this area as an emergent and prospective field of study.

Chapter 5 - According to FAO fisheries statistics, global shrimp production has shown an upward trend. Shrimp stand as an essential aspect of coastal fisheries for many countries worldwide. With climatic factors like fluctuations in sea surface temperature, elevating sea levels, coastal inundations, eroding coastlines, oceanic heatwaves, and heightened ocean acidity, there's growing concern for regions spanning both coastal and open oceans. This is especially significant as these are the primary zones for a majority of aquaculture operations. This chapter aims to synthesize the available literature on the impacts of climate change on global shrimp disease using the Scientometric method. The authors first identified the top disease in shrimp aquaculture along with climate change elements, then extracted bibliometrics from the Web of Science. Then the authors used CiteSpace to assess trends and research focusing on the impacts of climate change on worldwide shrimp disease. To address gaps in this emerging field, the authors make two main recommendations: (i) Increased collaboration between countries to develop global solutions towards shrimp disease* in the face of climate change and (ii) Drawing from other disciplines outside of "Ecology, Earth, Marine," such as molecular biology, economics, and sustainability, would strengthen and provide additional insights on sustainable shrimp production in the world. Further qualitative discussion was provided along with the future research topic's direction of Sustainable Development Goals of Food Security and Climate Change.

Chapter 6 - Global climate warming exerts a profound influence on myriad oceanic and marine ecosystem processes. This edited collection offers a timely glimpse into contemporary trends and developments concerning the impact of climate change on marine environments and their resources. The series underscores the unique opportunities presented by scientific studies, from selecting pivotal research problems to shaping career paths and advancing within a particular domain. Furthermore, it is anticipated that global warming will directly affect the marine food chain. In summation, this book furnishes a foundational overview of national adaptation policies and practices in the face of climate change.

Acknowledgments

This book was supported by funds provided by USDA-NIFA Sustainable Agriculture Systems (Grant No. 2019-69012-29905; title of project: Empowering US Broiler Production for Transformation and Sustainability USDA-NIFA (Sustainable Agriculture Systems): No. 2019-69012-29905) to Professor Dr. Guillermo Téllez-Isaías (GTI) at Department of Poultry science, University of Arkansas, USA. Most of the chapters in this book is supported primarily by the Ministry of Higher Education Malaysia (MOHE), under the Long Term Research Grant program (LRGS/1/2020/UMT/01/1; LRGS UMT Vot No. 56040) entitled "Ocean Climate Change: Potential Risk, Impact and Adaptation Towards Marine and Coastal Ecosystem Services in Malaysia', with a sub-project entitled "Charting the Effects of Climate Change and Acidification through Marine Organism Physiological Responses" to Dr. Mohamad Nor Azra (MNA). MNA also supported by the Ministry of Natural Resources, Environment and Climate Change, Malaysia or also known as NRECC under the National Conservation Trust Fund for Natural Resources (NCTF), entitled "Quantifying Red Claw Crayfish Distribution and Invasiveness in Peninsular Malaysia for Societal Wellbeing and Environmental Sustainability" reference No. (KeTSA.BPP.(S)600-5/9/17(1) with the internal Vote. No. of 53511. MNA also receives support from the Sustainable Ocean Alliance (SOA) and the Environmental Defense Fund (EDF) in the USA for his inaugural fellowship on Leadership for Climate Resilient Fisheries (LCRF), and Research Center for Marine and Land Bioindustry, Earth Sciences and Maritime Organization, National Research and Innovation Agency (BRIN), Indonesia for his Visiting Researcher Program. Dr. Hidir Ariffin acknowledge Higher Institution Centre of Excellence-Institute of Tropical Aquaculture and Fisheries (HICOE-AKUATROP) for his postdoctoral research program (2023-2025). Editors also acknowledge current Director, Professor Dr. Yeong Yik Sung and Deputy Director, Assoc. Professor Dr. Tan Min Pau at Institute of Climate Adaptation and Marine Biotechnology, Universiti Malaysia Terengganu for their full support during the publication of the book.

Chapter 1

Introduction to Marine Ecosystems and Climate Change

Hidir Ariffin[1,*]
and Mohamad Nor Azra[2,3]
[1]Institute of Climate Adaptation and Marine Biotechnology (ICAMB), Universiti Malaysia Terengganu (UMT), Terengganu, Malaysia
[2]Research Center for Marine and Land Bioindustry, Earth Sciences and Maritime Organization, National Research and Innovation Agency (BRIN), Pemenang, West Nusa Tenggara, Indonesia
[3]Higher Institution Centre of Excellence (HICoE), Institute of Tropical Aquaculture and Fisheries (AKUATROP), Universiti Malaysia Terengganu, Terengganu, Malaysia

Abstract

Generally, climate change has been defined as "any change in climate over time, whether due to natural variability or as a result of human activity." There are many elements and terminologies that have been used to identify or categorize "climate change" such as "global warming," or "extreme event," or "anthropogenic changes," or "greenhouse effects," or "weather pattern," or "sea level rise," or "coastal erosion," or "heat wave," or "acidification," or, "drought," or "flood," or "heavy rainfall," or "melt of the glacier," "hypoxia," or "harmful alga bloom," or "eutrophication," or "atmospheric pollution," or "permafrost," or "agriculture runoff" or "burning of fossil fuel." Meanwhile, the marine ecosystem is among one of the most impacted by the continuous changes in climate and concurrently receiving numerous stressors from natural and anthropogenic sources. However, limited studies of unequivocal theoretical framework on how marine life

---

* Corresponding Author's Email: hidirariffin@umt.edu.my.

In: Marine Life in Changing Climates
Editors: M. Nor Azra, T. Chandra Segaran, G. Téllez-Isaías et al.
ISBN: 979-8-89113-404-1

responds to climate change strongly compromises recent ability to make further predictions in the future. Thus, combining all of this knowledge on how marine life responds to climate change within the edited series is vital.

Keywords: marine, ecosystem services, climate change, environmental sciences, ocean

1. Introduction: Why Is Marine Life Important?

From the smallest plankton to the largest whale, all these organisms are included as marine life, which set up a complex ecosystem in the sea with millions of animal and plant species still undiscovered. To date, approximately 242,000 marine species have been cataloged, and as research progresses, the discovery of new species is anticipated to persist (Bouchet et al., 2023). Each organism contributes uniquely, culminating in a rich and invaluable marine ecosystem. One of the highlight marine life creatures is phytoplankton. These tiny drifters are referred to as plants and undergo photosynthesis for energy, ultimately producing 50-80 percent of global oxygen (Mondal and Manuel De la Sen, 2022). Without the biggest oxygen producers, it certainly harms every life in many parts of the world. In that regard, the ocean is considered the lung of the planet and that is why scientists exert the utmost effort to save marine life from the far-reaching consequences of climate change.

Marine ecosystems boast a diverse array of marine life, including fish, squid, mollusks, and crustaceans, which serve as primary protein sources for human consumption. Globally, seafood accounts for 17% of protein intake, and in developing nations, it's even more crucial, providing the primary protein source for over half of their populations (FAO, 2020). Through capture fisheries, a substantial 7,135 Kt of protein is contributed to the global supply, underscoring its importance for ensuring food security (Boyd and Davis, 2022).

2. How Is Marine Life Impacted by Climate Change?

In the past, the world demonstrated a series of climate changes from time to time however, the revitalization of human activities further expedites climate change up to a certain point that never the world experienced earlier.

Fluctuating atmospheric carbon dioxide (CO_2) has been observed within 180 ppm to 300 ppm over the last 650 000 years (Lamon and Marcomini, 2009). In the modern era, scientists verified the CO_2 concentration revved up at the quickest pace from 280 ppm to 379 ppm in 2005 (Lamon and Marcomini, 2009).

An increase in the global average surface temperature by 0.76°C over the century from 1900 to 2000. As a consequence, marine ecosystems bear a significant burden due to these human-induced temperature changes, often likened to experiencing a prolonged heat wave (Lamon and Marcomini, 2009). Climate model projections suggest that by 2100, up to 50% of all marine species could face extinction due to escalating sea temperatures, given that the vast expanses of seawater absorb nearly 80% of the global heat (Johnson and Welch, 2010). A pressing concern associated with these temperature increases is the stratification of seawater. This stratification, induced by elevated temperatures, can reduce oxygen concentrations, posing a significant threat to marine life (Hobday and Lough, 2011). Furthermore, ocean acidification is one of the prominent emerging marine life threats in which by 2100, the oceanic pH is anticipated to fall from 0.14 to 0.35 units (Kroeker and Gattuso, 2013). The excess anthropogenic atmospheric carbon dioxide sinks and dissolves into the ocean, forming weak carbonic acid, leading to a drop in seawater pH.

Consequently, marine life, especially marine calcifying organisms, faces urgent threats due to declining pH levels which compromise the integrity of carbonate structures. Additionally, a decrease in pH adversely affects numerous marine species such as corals, mollusks, and certain species of plankton, impacting both their growth and reproductive capabilities (Johnson and Welch, 2010).

3. Overview of the Book's Purpose and Scope

Researchers from around the globe are tirelessly working to address the challenges posed by climate change, continuously expanding and intensifying their investigations into its various facets. This dedication has resulted in a remarkable number of scientific articles on the topic. These publications are then systematically analyzed and synthesized, leading to the production of scientometric review papers.

Scientometric analysis offers a comprehensive examination of extensive research articles on a specific topic, in this case, climate change. This quantitative study is executed using specialized databases and software tools. Insights from these reviews reveal crucial information about key contributors, collaboration patterns, and emerging themes in the field. These insights, in turn, provide valuable direction for the scientific community, highlighting potential areas of research and addressing gaps.

To conduct a robust scientometric analysis, the accurate selection of keywords is paramount. In this book, we have focused on several keywords associated with climate change, including: (1) climate adaptation, (2) landing and captured fisheries, (3) molluscs, and (4) shrimp disease. Essentially, this book delves into the interrelationship between climate change and various domains like climate adaptation, fisheries, molluscs, and shrimp diseases. As scientometric analyses grow in popularity, they equip scientists with the tools to tailor their research to contemporary requirements. This book aims to encapsulate the current trajectory of research on climate change, thereby offering researchers precise areas of exploration for the future.

Acknowledgments

The present work was supported by the Long-Term Research Grant Scheme (LRGS) awarded by the Department of Higher Education, Ministry of Higher Education Malaysia (LRGS/1/2020/UMT/01/1; LRGS UMT Vot No. 56040) (2020—2024).

Funding

The edited chapter has been sponsored by USDA-NIFA Sustainable Agriculture Systems, Grant No. 2019-69012-29905 to one of the Editors, Prof. Dr. Guillermo Téllez-Isaías. Title of the project: Empowering US Broiler Production for Transformation and Sustainability USDA-NIFA (Sustainable Agriculture Systems): No. 2019-69012-29905.

References

Bouchet, P., Decock, W., Lonneville, B., B. Vanhoorne, & L. Vandepitte. (2023). Marine biodiversity discovery: the metrics of new species descriptions. *Frontiers in Marine Science*, *10*. https://doi.org/10.3389/fmars.2023.929989.

Boyd, C. E., McNevin, A. A., & Davis, R. P. (2022). The contribution of fisheries and aquaculture to the global protein supply. *Food Security*. https://doi.org/10.1007/s12571-021-01246-9.

FAO. (2020c). *The state of world fisheries and aquaculture*. Rome: FAO. http://www.fao.org/documents/card/en/c/ca9229en.

Hobday, A. J., & Lough, J. M. (2011). Projected climate change in Australian marine and freshwater environments. *Marine and Freshwater Research*, *62*(9), 1000. https://doi.org/10.1071/mf10302.

Johnson, J. E., & Welch, D. J. (2009). Marine Fisheries Management in a Changing Climate: A Review of Vulnerability and Future Options. *Reviews in Fisheries Science*, *18*(1), 106–124. https://doi.org/10.1080/10641260903434557.

Kroeker, K. J., Kordas, R. L., Crim, R., Hendriks, I. E., Ramajo, L., Singh, G. S., Duarte, C. M., & Gattuso, J.-P. (2013). Impacts of ocean acidification on marine organisms: quantifying sensitivities and interaction with warming. *Global Change Biology*, *19*(6), 1884–1896. https://doi.org/10.1111/gcb.12179.

Lamon, L., Dalla Valle, M., Critto, A., & Marcomini, A. (2009). Introducing an integrated climate change perspective in POPs modelling, monitoring and regulation. *Environmental Pollution*, *157*(7), 1971–1980. https://doi.org/10.1016/j.envpol.2009.02.016.

Mondal, S., Samanta, G., & De la Sen, M. (2022). Dynamics of Oxygen-Plankton Model with Variable Zooplankton Search Rate in Deterministic and Fluctuating Environments. *Mathematics*, *10*(10), 1641. https://doi.org/10.3390/math10101641.

Chapter 2

Environmental Adaptations to Climate Change: A Scientometric Analysis

Nur Asniza Aziz[1,*,+]
Mohamad Nor Azra[1,2,†,+]
Thirukanthan Chandra Segaran[1]
Mohd Iqbal Mohd Noor[3,4]
Yeong Yik Sung[2]
Mahmoud A. O. Dawood[5,6]
Mazlan Abd Ghaffar[2,7,8]
Tan Min Pau[1]
and Walter G. Bottje[9]

[1]Institute of Climate Adaptation and Marine Biotechnology (ICAMB), Universiti Malaysia Terengganu (UMT), Terengganu, Malaysia
[2]Research Center for Marine and Land Bioindustry, Earth Sciences and Maritime Organization, National Research and Innovation Agency (BRIN), Pemenang, West Nusa Tenggara, Indonesia
[3]Faculty of Business Management, Universiti Teknologi MARA (UiTM) (Pahang), Raub, Pahang, Malaysia
[4]Institute for Biodiversity and Sustainable Development, Universiti Teknologi MARA (UiTM), Shah Alam, Selangor, Malaysia
[5]The Center for Applied Research on the Environment and Sustainability, The American University in Cairo, Cairo, Egypt
[6]Animal Production Department, Faculty of Agriculture, Kafrelsheikh University, Kafr El-Sheikh, Egypt

* Corresponding Author's Email: asniza.aziz@umt.edu.my.
† Corresponding Author's Email: azramn@umt.edu.my.

In: Marine Life in Changing Climates
Editors: M. Nor Azra, T. Chandra Segaran, G. Téllez-Isaías et al.
ISBN: 979-8-89113-404-1

[7]Faculty of Science and Marine Environment, Universiti Malaysia Terengganu, Terengganu, Malaysia
[8]Higher Institution Centre of Excellence (HICoE), Institute of Tropical Aquaculture and Fisheries (AKUATROP), Universiti Malaysia Terengganu, Terengganu, Malaysia
[9]Department of Poultry Science, University of Arkansas, Fayetteville, AR, USA
[+]Authors sharing the first co-authorship

Abstract

As global climate change intensifies, both human and natural ecosystems must adapt to its multifaceted impacts. Extensive research has already been conducted to gauge how these environments respond to and prepare for climate-induced alterations. Collating this knowledge is essential, as it not only offers a consolidated view of present findings but also highlights areas yet to be explored. To this end, we surveyed pertinent English-language articles from the Web of Science Core Collection (WOSCC) using keywords "climate adaptation" and "climate change adaptation." Utilizing the CiteSpace software, we constructed visual knowledge maps to evaluate the present state and developmental trajectory of climate change adaptation research. From our search, we amassed 9,098 articles in the WOSCC, published between 1972 and 31-12-2021. Notably, around 53% of these were contributions from the USA, Australia, and England. Over the past five decades, this field has seen participation from 6,664 institutions, collaboration among 26,814 authors, and publications across 1,680 distinct journals. Through CiteSpace, we discerned 14 primary clusters, with prominent themes including spatial planning, nature-based solutions, and strategies for smallholder farmers. Recurring keywords encompassed "vulnerability," "management," "evolution," "geographic variation," "body size," and "temperature." The emphasis on restructuring human habitats via spatial planning and nature-based solutions underscores the scientific community's proactive approach to climate change adaptation. A significant portion of this work pertains to the realms of "Ecology," "Earth & Marine," and "Economics, Economic & Political Studies." Our findings advocate for policymakers to prioritize localized adaptation strategies and foster informed decision-making to adeptly navigate the challenges of climate change. This chapter serves as a valuable resource for scientific communities, philanthropic entities, relevant government bodies, and NGOs, guiding them in the sustainable management of the repercussions of climate change.

Keywords: scientometric, evolution, body size, geographic variation, temperature

1. Introduction

The concept of adaptation fundamentally encompasses adjustments made by humans or the environment in response to changing climatic conditions or their effects. Such adjustments can serve to mitigate harm or seize opportunities (Orlove, 2005; FAO, 2022; Rivera-Collazo, 2022). Broadly speaking, there are three primary pillars in addressing climate change: mitigation, adaptation, and resilience. All three are of equal importance in combating climate change (Descheemaeker et al. 2016; Castells-Quintana et al. 2018; Einecker and Kirby, 2020; Geiger et al. 2021). However, due to the longevity of greenhouse gases in the Earth's atmosphere, their effects are inevitable (Ledley et al. 1999; Solomon et al. 2009; Kocak and Alnour, 2022). Human behavior plays a pivotal role in shaping sustainable energy and climate policies (Makijenko et al. 2016). While the natural environment can adapt through human intervention, comprehensive strategies to prepare for these climatic changes are often underrepresented in scientific publications, especially in the realm of scientometric analysis. This gap underscores the primary motivation for this chapter: to spotlight adaptation strategies to climate change, a topic gaining prominence globally.

Climate change, marked by increasing temperatures, rising sea levels, and prolonged droughts, is anticipated to significantly impact biodiversity, agricultural output, and food security. Concurrently, many studies have noted alterations in the intensity and magnitude of the dual monsoon seasons. The frequency of extreme weather events, characterized by powerful winds, torrential rains, towering waves, and extended droughts, is on the rise. Events such as the 2004 tsunami and the massive floods of 2006 and 2014 stand as testament to their devastating power, affecting numerous nations and ecosystems. The economic aftermath of such events is staggering, often reaching into the billions. The tangible consequences, including loss of life, property damage, and disruption to livelihoods, are immediate challenges communities face. Research focusing on the implications of climate change on sea level, especially in tropical regions, has been undertaken to forecast coastal conditions and potential sea level rise by the year 2100.

It showed that there is a significant increase in sea level rise trend over the last 5 years, compared to 20 years ago. Impacts and adaptation measures climate change and sea level rise can give rise to high impacts such as destruction of assets and economic sectors, loss of human lives, mental health effects, loss on plants, animals, and ecosystem and their severity depends on their extremes, exposure and vulnerability (IPCC, 2012). Sea level rise may also reduce the size of an island and its infrastructure and thus, compromise the socio-economic wellbeing of the island communities (Handmer et al., 2012). These impacts of disasters may be minimized or avoided with knowledge and preparedness.

A definition of scientometric is "systematic, continuous and comprehensive assessment of quantitative study on patterns of science, technology and innovation" (Mooghali et al. 2011; Abramo, 2018; Kim Zhu, 2018). Scientometric analysis or also known as macroanalysis of scientific publication is considered as an important review and extensively covers and reviews the number of works by different prominent researchers and publications on the subject through scientific database and technical software (Borner et al. 2003; Azzeri et al. 2020; Noor et al. 2021; Azra et al. 2022). There is a wealth of scientific publication that are attempts to understand how both environments adapt to climate change adaptation. Thus, synthesizing this evidence is vital as it provides a critical perspective on current results, and insights on gaps for future research. This chapter aim to identify the development trends of climate change adaptation through visualized knowledge map of (i) evolution of publication in terms of number of scientific papers, open access option and countries distribution (i) co-citation analysis of author, countries and document (ii) cluster analysis of published articles and (ii) dual map overlay as well as (iv) keywords burstness.

Knowledge of climate change is a rising matter invoking interest from many stakeholders because of it will impacted various environmental and ecosystem services. Ecosystems provide essential goods and services supporting human health, livelihoods, wellbeing and survival, these have been termed ecosystem services (ES). Several categories of climate change are about how the climate system works; specific knowledge about the causes, consequences and possible solutions and practical knowledge for individual and collective actions. However current knowledge of global climate change and its associated environmental issues have received little attention globally, especially on the climate adaptation. Shifts in marine environments, which have consequential effects on planetary health, underscore the importance of understanding climate change and its manifold impacts. Every nation across all continents is grappling with the repercussions of climate change, witnessing disruptions in their economies and daily lives. Altered weather patterns, escalating sea levels, and intensifying weather events are becoming the new norm. An analysis by Climate Central suggests that certain coastal regions could be inundated by 2030 due to these rising sea levels. Nonetheless, several of these countries remain relatively uninformed about the profound effects of climate change, primarily because its impacts might be indirectly experienced in their daily lives. A comprehensive understanding of climate change, bolstered by rigorous research, can spark genuine conviction, subsequently catalyzing positive behavioral shifts within the general populace (Abas et al., 2017).

This chapter delves into a vital aspect of climate change, encompassing a range of crucial topics. By focusing on specific keywords related to the subject, we highlight seminal research works from leading documents in the field. This review not only furnishes insights into existing knowledge gaps, directing future research endeavors, but also amalgamates the current body of evidence sourced from scientific databases. Such a synthesis aims to inform the formulation of adept adaptation strategies to manage the repercussions of climate change. Furthermore, this review is a component of a broader research initiative striving to protect both human and environmental resources globally.

2. Methods

In the methodological section, we explain the procedure for analyzing the scientific in climate adaptation research. We describe the secondary data that we used through macroanalysis of descriptive and citation approaches.

2.1. Data Sources

Web of Science database of Core Collection or also known as WOSCC administered by international organization of Thomson Reuters of Scientific Information was the only database used in this chapter.

The Web of Science (WOS) stands as one of the most expansive and detailed citation databases for scrutinizing peer-reviewed scientific endeavors, including bibliometric or scientometric studies (Adriaanse and Rensleigh, 2013; Aryadoust and Ang, 2019). For the scope of climate change adaptation within WOSCC, the search parameters employed were "climat* adapt*" and "climat* chang* adapt*" under the Topic Selection (TS) category. We confined our search to original research articles, spanning the inception of WOSCC up to the conclusion of December 2021. The analysis was further narrowed to encompass only articles written in English. Data extracted from WOSCC were archived as plain text files (.txt), each encompassing a maximum of 500 records—which included the title, abstract, keywords, and cited references.

2.2. Scientometric Analysis

Visual knowledge maps was generated by CiteSpace software, in which it was developed to analyze the current body of research and to assess development trends of selected knowledge domain (Chen, 2006). The specific version utilized, CiteSpace version 6.1.2 (64-bit for Windows), was developed by Chaomei Chen in 2016 and has since been adopted by researchers worldwide (Chen, 2020). The software operates based on a distinctive workflow, encompassing a sequence of analytical and integrative tasks, and offers an extensive range of strategy and operation combinations. Within the CiteSpace environment, terms like "burst detection," "centrality," "sigma," and "cluster analysis" are frequently encountered (Chen, 2020). Meanwhile, the co-citation is the common phrase used in the scientometric analysis. Co-cited documents define as two references are cited by the third article, and analyzing such networks will benefited for identify patterns and trends of underlying scientific publication.

3. Results

3.1. Trends in Scientific Publications

Figure 1 indicates the research on climate change adaptation grew rapidly between 2018 and 2021. A notable data point from 2021 shows that 1,520 papers were cataloged in the WOSCC database—a tenfold increase from the number in 2010. This uptrend in article publication underscores the escalating interest in climate change adaptation over the past decade. Interestingly, nearly 50% of these articles are accessible under the open access category, with a substantial portion also available through the "free to read" option, as depicted in Figure 2. Geographical analysis, shown in Figure 3, reveals that the USA, Australia, and England collectively contributed to about 53% of the total publications. A retrospective look over the past five decades highlights that 6,664 institutions, 26,814 distinct authors, and 1,680 diverse journals have actively engaged in the research and development of climate change adaptation (refer to Supplementary Materials for more details).

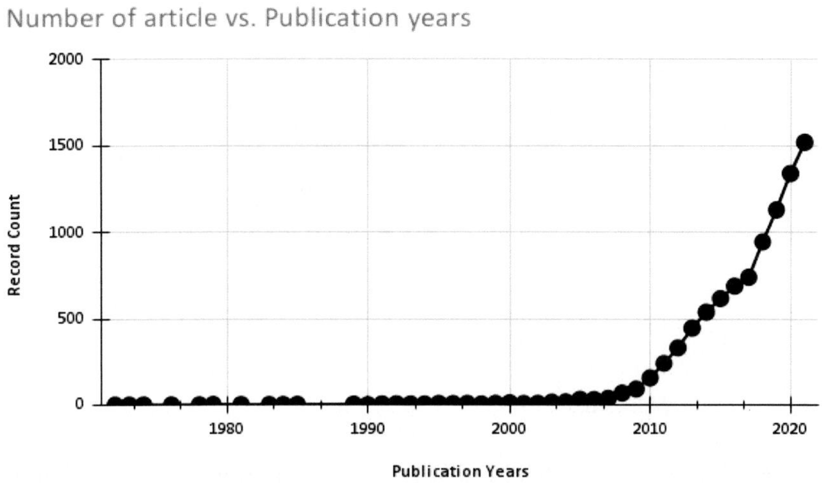

Figure 1. Publication Trends in Climate Change Adaptation within the Web of Science Core Collection.

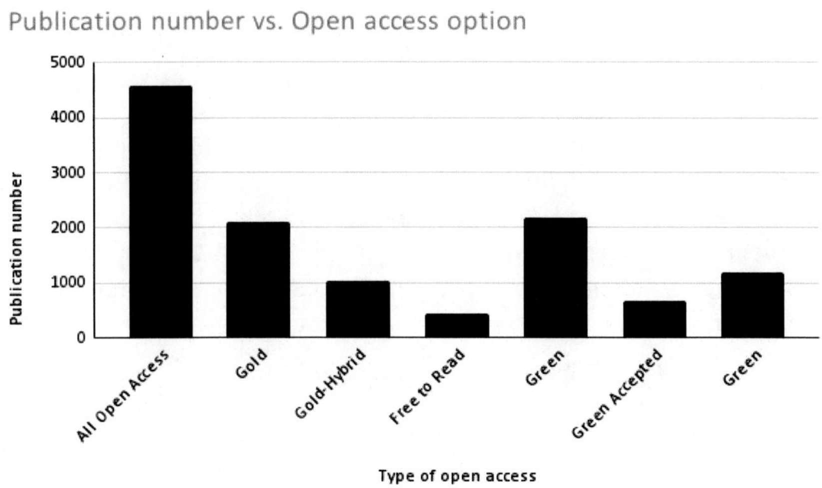

Figure 2. Distribution of Publications by Access Type in Climate Change Adaptation Research.

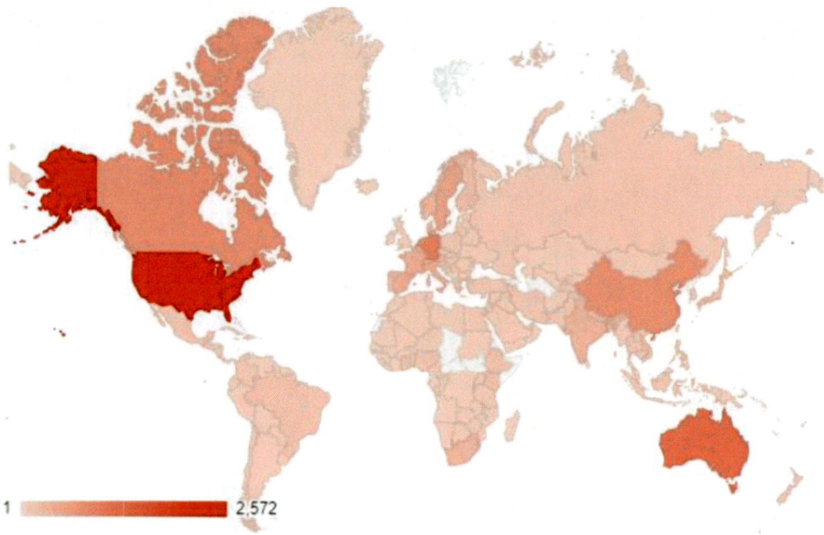

Figure 3. Geographic Distribution of Articles on Climate Change Adaptation in the Web of Science Core Collection.

3.2. Analysis of Citation and Its Networks

In this chapter, there are four different types of scientometric data generated from the CiteSpace software, which are (i) co-citation analysis (ii) cluster analysis, (ii) dual map overlay and (iv) keywords burstness. The co-cited analysis was divided into three different co-citation which is collaborating countries (Figure 4), co-cited authors (Figure 5) and (iii) co-cited documents (Figure 6). This chapter also showed that the network analysis is divided into 14 co-citation clusters (Figure 7).

The primary discipline in climate change adaptation research is "Ecology, Earth, Marine." This is succeeded by four distinct fields, namely:

1. "Earth, Geology, Geophysics"
2. "Plant, Ecology, Zoology"
3. "Economics, Economic, Political"
4. "Molecular, Biology, Genetics"

The interconnectivity between these fields and disciplines is visually represented in the Dual Map Overlay analysis (refer to Figure 8). Burst analysis of keywords reveals that "climatic adaptation," "evolution," and "geographic variation" are the three most emergent terms in the realm of climate change adaptation (see Table 1). Moreover, when focusing on the Topic selection within the Web of Science Core Collection, the predominant keywords in climate change adaptation research include “vulnerability,” “management,” “evolution,” “geographic variation,” “body size,” and “temperature” (as illustrated in Figure 9).

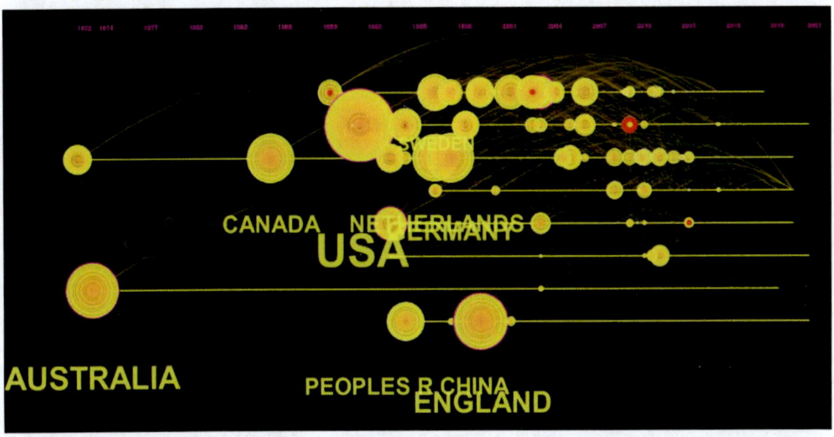

Figure 4. Co-citation Analysis of Collaborative Countries in Climate Change Adaptation Research.Co-citation of collaborating countries in the field of climate change adaptation.

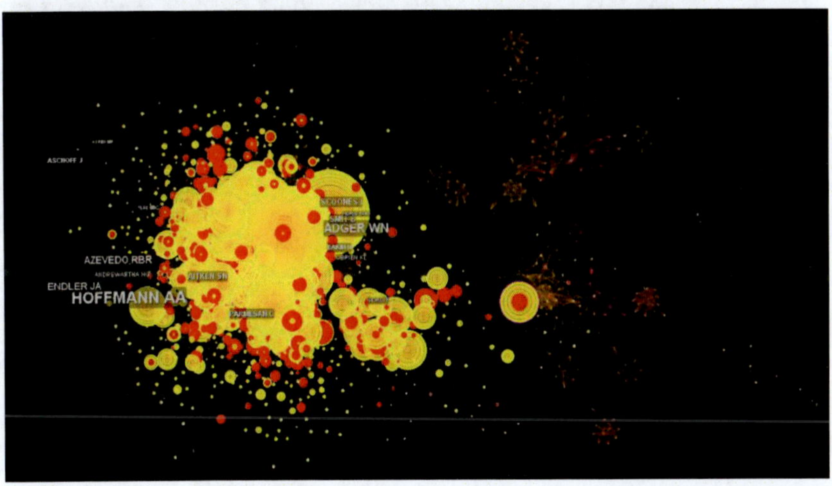

Figure 5. Network of Co-cited Authors in Climate Change Adaptation Research.

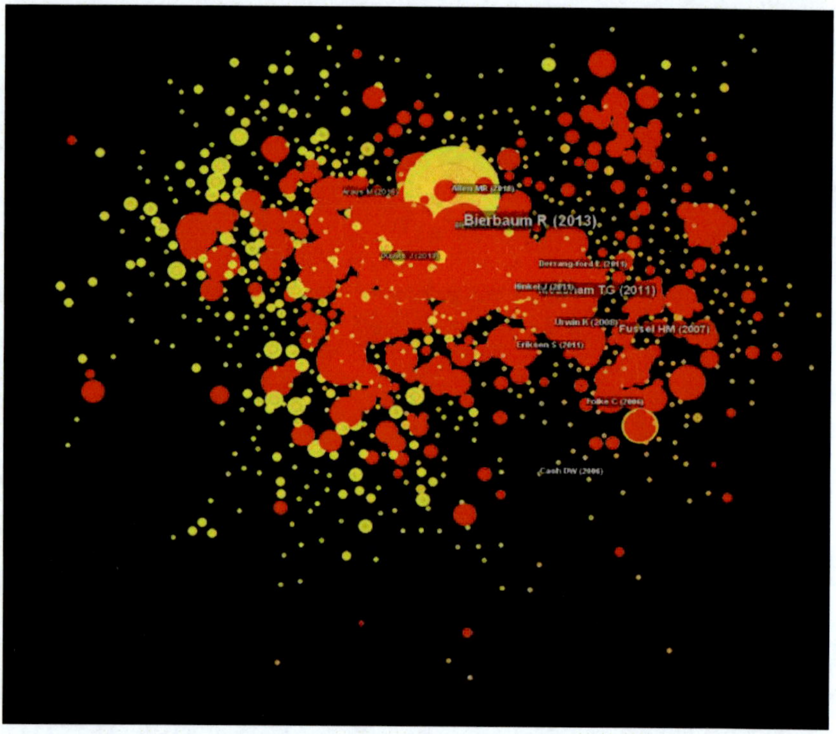

Figure 6. Document Networks for Co-citation Analysis in the Realm of Climate Change Adaptation.

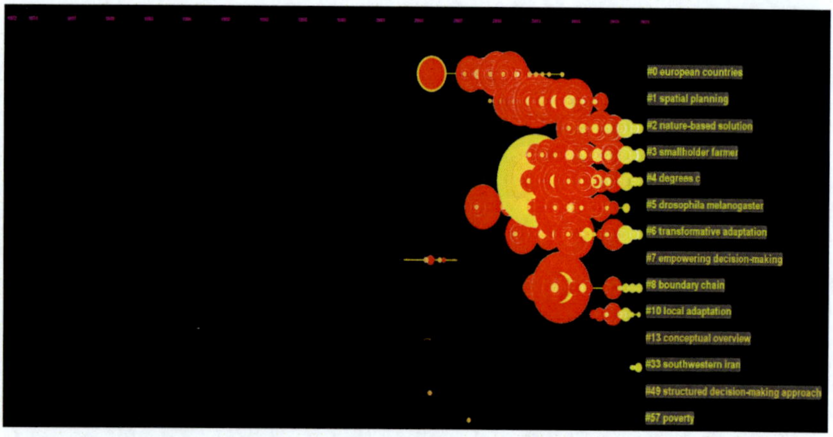

Figure 7. Cluster Analysis Visualization for Climate Change Adaptation using CiteSpace Software.

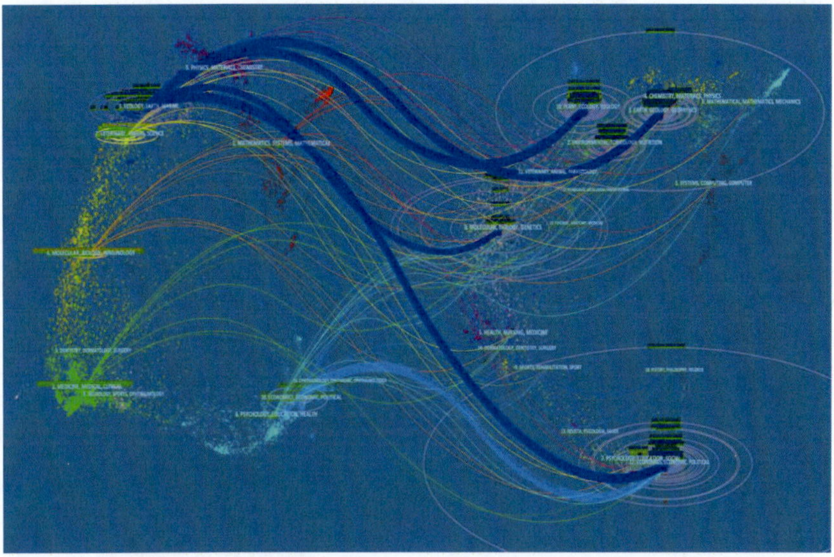

Figure 8. Dual Map Overlay for Climate Change Adaptation Research, Based on Original Articles and Citation Data.

Figure 9. Distribution of Keywords in Climate Change Adaptation Research from the Web of Science Core Collection.

Table 1. Top 10 Keywords with the Strongest Citation Bursts in terms of climate change adaptation research

Keywords	Year	Strength	Begin	End	1972 - 2021
climatic adaptation	1972	47.01	1991	2012	
evolution	1972	17.83	1995	2012	
geographic variation	1972	11.67	1994	2016	
body size	1972	9.46	1992	2008	
temperature	1972	8.43	1999	2009	
natural population	1972	8.2	1995	2010	
population	1972	7.8	1993	2010	
melanogaster	1972	6.7	1993	2012	
drosophila melanogaster	1972	5.98	1995	2012	
selection	1972	4.68	1998	2010	

4. Discussion

Tropical regions, characterized by their equatorial location, have long been a focal point of climate studies due to their unique challenges. Being directly under the sun's path and being perpendicular to the earth's rotation axis at the equator, these regions experience intense sunlight and warmth. However, the challenges these regions face have been further compounded by the repercussions of climate change (CC). The greenhouse effect, driven largely by increased concentrations of greenhouse gases like carbon dioxide (CO_2), leads to global temperature rise, subsequently causing sea-level rise.

Internationally, climatologists and researchers are fervently studying the earth's climate system in its entirety. Their findings and models inform the creation of global policies that countries can adopt to mitigate the impacts of climate change. On a more localized level, the pressing need to understand the specific impacts of climate change has catalyzed efforts to develop adaptation plans tailored to regional and local climates, especially in tropical regions.

Organizations like the United Nations (UN) have defined climate change as a long-term alteration in typical weather patterns in a place. Corroborating this, the National Aeronautics and Space Administration (NASA) added that shifts may manifest as changes in a region's usual rainfall amounts or temperature patterns during specific months or seasons. NASA's research further posits that we are on a trajectory where the earth's average temperatures will continue to rise for the next century. A direct consequence of this is the ocean's absorption of excess heat, resulting in sea-level rise (SLR).

Undoubtedly, global warming stands out as a pressing environmental concern. The primary culprit exacerbating this issue is carbon dioxide (CO2), a major greenhouse gas in our atmosphere. Its role in trapping heat by absorbing and re-emitting infrared light is central to the global temperature increase. Forecasts by the International Panel on Climate Change (IPCC) predict a somber future, suggesting that atmospheric CO2 concentrations could soar to 590 ppm by 2100, causing a global average temperature spike of 1.9°C. Such a scenario will have vast, global ramifications. These include polar ice melt, accelerated sea-level rise, and a surge in global precipitation (Khatib, 2012). Notably, the primary contributors to CO2 emissions are the energy sector, especially fossil fuel combustion, and agriculture.

Climate change adaptation has become increasingly attractive field in the climate science research. Australia is among the first countries that are actively involved in the climate change adaptation research, followed by Canada and USA. Among the nations collaborating in the field of climate change adaptation, the USA, England, and South Africa are frequently cited. Ary A. Hoffmann from The University of Melbourne, Australia stands out as a particularly influential author within this knowledge domain. His prominence is evidenced by a high sigma score, as depicted by the larger word size in Figure 5.

Bierbaum et al. (2013) have produced one of the most influential documents in climate change adaptation research, as illustrated in Figure 6. In their seminal study, Bierbaum and his team explored existing adaptation plans and related activities at the local level. Their findings highlighted the pivotal roles of both the public and private sectors in formulating substantial planning for climate change adaptation. The team emphasized the necessity for robust measures to safeguard livelihoods, as well as the use of land and water resources.

We identified 14 clusters from CiteSpace software, including spatial planning, nature-based solutions, and smallholder farmers. Cluster such as "European countries," "spatial planning," "nature-based solution," "smallholder farmer" is among the top four cluster in the field of climate change adaptation. The largest cluster (#0) has 242 members (i.e., article published) and was labeled as european countries. The most relevant citer to the cluster is Wise et al. (2014) entitled "Reconceptualising adaptation to climate change as part of pathways of change and response," published by Global Environmental Change journal. They introduced the pathway approach for future guidance to policy makers as well as the creation of mechanisms for further funding in climate change adaptation.

This indicates that primary research clusters are concentrating on reshaping present human environments through spatial planning and nature-based solutions. Such a trend suggests that the scientific community is actively pursuing strategies to adapt to climate change. Interestingly, based on the dual map overlay analysis, generated from the CiteSpace software, there is only a single field that was linked in the climate change adaptation research, which is "Ecology, Earth, Marine." The most current keywords burstness is tagged as "geographic variation."

This underscores the notion that the significance of climate change research can vary based on geographic location. Interestingly, "Drosophila melanogaster" ranks among the top 10 influential keywords in the climate change adaptation literature. This suggests that this species has been extensively employed as a model organism in studies aiming to understand and facilitate climate change adaptation, as highlighted by works such as Rane et al. (2015) and Rodrigues and Cogni (2021).

Conclusion

This chapter providing insights from a diverse range of perspectives and sharing experienced for researchers who are new in the field of climate change adaptation. Much of this research focused on the field of "Ecology," "Earth & Marine" as well as "Economics, Economic & Political." Our results suggest that policy makers should focus on local adaptation and empowering decision making in order to properly adapt to climate change. This chapter benefits scientific communities, philanthropic funders, related governments, and

NGOs towards sustainably managing ongoing climate change impacts. The public must be made aware and be informed about climate change consequences and the actions they could adopt to adapt to climate change. Finally, the result of this chapter hopefully would help them in developing public awareness of climate change impact and also determining the best adaptation measures to climate change to be adopted by island communities in the world. Recently, the United Nations Development Programme's Sustainable Development Goals (SDG) for Malaysia, Thailand, and Singapore outlined their ambitions for achieving a sustainable future. Goal 13 emphasizes the imperative of climate action. In alignment with the objectives of our chapter—which aims to introduce new methodologies for discerning contemporary knowledge on addressing climate change—the implications for society, academia, governmental agencies, industry, and the environment are clear and direct.

Acknowledgments

This chapter have been presented in the Inaugural Conference co-hosted by Institute of Marine Biotechnology, currently known as Institute of Climate Adaptation and Marine Biotechnology (ICAMB), Universiti Malaysia Terengganu, Terengganu, Malaysia Ghent University, Belgium, entitled "International Conference of Marine Biotechnology (i-CoMB) 2022 "as invited speaker presentation. This chapter is supported by the Department of Higher Education, Ministry of Higher Education Malaysia under the LRGS program (LRGS/1/2020/UMT/01/1; LRGS UMT Vot No. 56040) entitled 'Ocean climate change: potential risk, impact and adaptation towards marine and coastal ecosystem services in Malaysia'. We also acknowledge the funds provided by USDA-NIFA Sustainable Agriculture Systems, Grant No. 2019-69012-29905. Title of Project: Empowering US Broiler Production for Transformation and Sustainability USDA-NIFA (Sustainable Agriculture Systems): No. 2019-69012-29905.

Funding

The coloured figures for the present chapter has been sponsored by USDA-NIFA Sustainable Agriculture Systems, Grant No. 2019-69012-29905 to one of the Editors, Prof. Dr. Guillermo Téllez-Isaías. Title of the project: Empowering US Broiler Production for Transformation and Sustainability USDA-NIFA (Sustainable Agriculture Systems): No. 2019-69012-29905.

References

Abas, N., Daud, Z. M., Mohamed, N., & Halim, S. A. (2017). Climate change impact on coastal communities in Malaysia. *Journal of Advanced Research Design* 33: 1-7. https://www.researchgate.net/publication/319097200_Climate_change_impact_on_coastal_communities_in_Malaysia.

Abramo, G. (2018). Revisiting the scientometric conceptualization of impact and its measurement. *Journal of Informetrics*, 12(3), 590–597. https://doi.org/10.1016/j.joi.2018.05.001.

Adriaanse, L., & Rensleigh, C. (2013). Web of Science, Scopus and Google Scholar. *The Electronic Library*, 31(6), 727–744. https://doi.org/10.1108/el-12-2011-0174.

Aryadoust, V., & Ang, B. H. (2019). Exploring the frontiers of eye tracking research in language studies: a novel co-citation scientometric review. *Computer Assisted Language Learning,* 34(7), 898–933. https://doi.org/10.1080/09588221.2019.1647251.

Azra, M. N., Jye, M. W., Van Doan, H., Zekker, I., Abdullah, M., Latif, Z. A., & Noor, M. I. M. (2022). Mapping of marine lobster research: A global outlook. *Frontiers in Marine Science*, 9. https://doi.org/10.3389/fmars.2022.976199.

Azzeri, A., Goh, H. C., Jaafar, H., Noor, M. I. M., Razi, N. A., Then, A. Y., Suhaimi, J., Kari, F., & Dahlui, M. (2020). A review of published literature regarding health issues of coastal communities in Sabah, Malaysia. *International Journal of Environmental Research and Public Health*, 17(5), 1533. https://doi.org/10.3390/ijerph17051533.

Bierbaum, R., Smith, J. B., Lee, A., Blair, M., Carter, L., Chapin, F. S., Fleming, P., Ruffo, S., Stults, M., McNeeley, S. M., Wasley, E., & Verduzco, L. (2012). A comprehensive review of climate adaptation in the United States: more than before, but less than needed. *Mitigation and Adaptation Strategies for Global Change*, 18(3), 361–406. https://doi.org/10.1007/s11027-012-9423-1.

Börner, K., Chen, C., & Boyack, K. W. (2005). Visualizing knowledge domains. *Annual Review of Information Science and Technology*, 37(1),179–255. https://doi.org/10.1002/aris.1440370106.

Castells-Quintana, D., Del Pilar López-Uribe, M., & McDermott, T. K. J. (2018). Adaptation to climate change: A review through a development economics lens. *World Development*, 104, 183–196. https://doi.org/10.1016/j.worlddev.2017.11.016.

Chen, C. (2006). CiteSpace II: Detecting and visualizing emerging trends and transient patterns in scientific literature. *Journal of the Association for Information Science and Technology*, 57(3), 359–377. https://doi.org/10.1002/asi.20317.

Chen C. (2020). *How to Use CiteSpace*. Lean Publishing British Columbia Canada p 246. https://citespace.podia.com/view/downloads/ebook-how-to-use-citespace.

Descheemaeker, K., Oosting, S., Tui, S. H., Masikati, P., Falconnier, G. N., & Giller, K. E. (2016). Climate change adaptation and mitigation in smallholder crop–livestock systems in sub-Saharan Africa: a call for integrated impact assessments. *Regional Environmental Change*, 16(8), 2331–2343. https://doi.org/10.1007/s10113-016-0957-8.

Einecker, R., & Kirby, A. (2020). Climate Change: A Bibliometric study of adaptation, Mitigation and Resilience. *Sustainability*, 12(17), 6935. https://doi.org/10.3390/su12176935.

FAO, (2022). FAO Terminology. *Climate change terminology based on Environmental Science*. Food and Agriculture Organization of the United Nations. (n.d.). https://www.fao.org/faoterm/en/.

Geiger, N., Swim, J. K., & Benson, L. (2021). Using the three-pillar model of sustainability to understand lay reactions to climate policy: A multilevel approach. *Environmental Science & Policy*, 126, 132–141. https://doi.org/10.1016/j.envsci.2021.09.023.

Handmer, J., Y. Honda, Z.W. Kundzewicz, N. Arnell, G. Benito, J. Hatfield, I.F. Mohamed, P. Peduzzi, S. Wu, B. Sherstyukov, K. Takahashi, and Z. Yan, 2012: *Changes in impacts of climate extremes: human systems and ecosystems*. In: Managing the Risks of Extreme Events and Disasters to Advance Climate Change Adaptation [Field, C. B., V. Barros, T. F. Stocker, D. Qin, D. J. Dokken, K. L. Ebi, M. D. Mastrandrea, K. J. Mach, G.-K. Plattner, S. K. Allen, M. Tignor, and P. M. Midgley (eds.)]. A Special Report of Working Groups I and II of the Intergovernmental Panel on Climate Change (IPCC). Cambridge University Press, Cambridge, UK, and New York, NY, USA, pp. 231-290. https://www.ipcc.ch/site/assets/uploads/2018/03/SREX-Chap4_FINAL-1.pdf.

IPCC (2012). IPCC, 2012 – *Managing The Risks Of Extreme Events And Disasters To Advance Climate Change Adaptation*, C. B., V. Barros, T. F. Stocker, D. Qin, D. J. Dokken, K. L. Ebi, M. D. Mastrandrea, K. J. Mach, G.-K. Plattner, S. K. Allen, M. Tignor, and P.M. Midgley (Eds.) Available from Cambridge University Press, The Edinburgh Building, Shaftesbury Road, Cambridge CB2 8RU ENGLAND, 582 pp. https://www.ipcc.ch/site/assets/uploads/2018/03/SREX_Full_Report-1.pdf.

Khatib, H. (2012). IEA World Energy Outlook 2011—A comment. *Energy Policy*. 48: 737-743. https://doi.org/10.1016/j.enpol.2012.06.007.

Kim, M. C., & Zhu, Y. (2018). *Scientometrics of Scientometrics: Mapping historical footprint and emerging technologies in scientometrics*. In InTech eBooks. https://doi.org/10.5772/intechopen.77951.

Koçak, E., & Alnour, M. (2022). Energy R&D expenditure, bioethanol consumption, and greenhouse gas emissions in the United States: Non-linear analysis and political implications. *Journal of Cleaner Production*, 374, 133887. https://doi.org/10.1016/j.jclepro.2022.133887.

Ledley, T. S., Sundquist, E. T., Schwartz, S. E., Hall, D. K., Fellows, J. D., & Killeen, T. L. (1999). Climate change and greenhouse gases. *EOS, Transactions American Geophysical Union*, 80(39), 453–458. https://doi.org/10.1029/99eo00325.

Makijenko, J., Burlakovs, J., Brizga, J., & Klavins, M. (2016). Energy efficiency and behavioral patterns in Latvia. *Management of Environmental Quality: An International Journal*, 27(6), 695–707. https://doi.org/10.1108/meq-05-2015-0103.

Mooghali, A., Alijani, R., Karami, N., & Khasseh, A. A. (2012). Scientometric analysis of the scientometric literature. *International Journal of Information Science and Management*, 9(1), 19–31. http://ijism.ricest.ac.ir/index.php/ijism/article/download/89/80.

Mohd Iqbal, M. N., Mohamad, N. A., Voon-Ching Lim, Azniza, A. Z., Dali, F., Hashim, I. M., . . . Abdullah, M. F. (2021). Aquaculture research in southeast asia - A scientometric analysis (1990-2019). *International Aquatic Research*, 13(4), 271-288. https://doi.org/10.22034/iar.2021.1932503.1166.

Orlove, B. S. (2005). Human adaptation to climate change: a review of three historical cases and some general perspectives. *Environmental Science & Policy*, 8(6), 589–600. https://doi.org/10.1016/j.envsci.2005.06.009.

Rane, R., Rako, L., Kapun, M., Lee, S. F., & Hoffmann, A. A. (2015). Genomic evidence for role of inversion3RPofDrosophila melanogasterin facilitating climate change adaptation. *Molecular Ecology*, 24(10), 2423–2432. https://doi.org/10.1111/mec.13161.

Rivera-Collazo, I. (2022). Environment, climate and people: Exploring human responses to climate change. *Journal of Anthropological Archaeology*, 68, 101460. https://doi.org/10.1016/j.jaa.2022.101460.

Rodrigues, M. F., & Cogni, R. (2021). Genomic responses to climate change: Making the most of the Drosophila model. *Frontiers in Genetics*, 12. https://doi.org/10.3389/fgene.2021.676218.

Solomon, S., Plattner, G., Knutti, R., & Friedlingstein, P. (2009). Irreversible climate change due to carbon dioxide emissions. *Proceedings of the National Academy of Sciences of the United States of America*, 106(6), 1704–1709. https://doi.org/10.1073/pnas.0812721106.

Wise, R., Fazey, I., Smith, M. S., Park, S. E., Eakin, H., Archer, E., & Campbell, B. M. (2014). Reconceptualising adaptation to climate change as part of pathways of change and response. *Global Environmental Change-human and Policy Dimensions*, 28, 325–336. https://doi.org/10.1016/j.gloenvcha.2013.12.002.

Chapter 3

Mapping the Research Landscape and Identifying Emerging Trends in Climate Change Impacts on Capture Fisheries and Fish Landing: A Scientometric Review

Thirukanthan Chandra Segaran[1,*]
Murni Nur Islamiah Kassim[1,†]
Nora Faten Afifah Mohamad[1]
Fathurrahman Lananan[3]
Youji Wang[4]
Guillermo Téllez-Isaías[5]
Walter G. Bottje[5]
Zulhisyam Abdul Kari[6]
and Mohamad Nor Azra[1]

[1]Climate Change Adaptation Laboratory, Institute of Climate Adaptation & Marine Biotechnology (ICAMB), Universiti Malaysia Terengganu (UMT), Kuala Nerus, Terengganu, Malaysia
[2]Research Center for Marine and Land Bioindustry, Earth Sciences and Maritime Organization, National Research and Innovation Agency (BRIN), Pemenang, West Nusa Tenggara, Indonesia
[3]East Coast Environmental Research Institute, Universiti Sultan Zainal Abidin (UniSZA), Kuala Terengganu, Malaysia
[4]International Research Center for Marine Biosciences at Shanghai Ocean University Ministry of Science and Technology, Shanghai, China
[5]Department of Poultry Science, University of Arkansas, Fayetteville, AR, USA

[*] Corresponding Author's E-mail: thiru@umt.edu.my.
[†] Corresponding Author's E-mail: murni.islamiah@umt.edu.my.

In: Marine Life in Changing Climates
Editors: M. Nor Azra, T. Chandra Segaran, G. Téllez-Isaías et al.
ISBN: 979-8-89113-404-1

[6]Department of Agricultural Science, Faculty of Agro-Based Industry, Universiti Malaysia Kelantan, Jeli, Kelantan, Malaysia

Abstract

Climate change poses significant challenges to the sustainability of capture fisheries, a vital component of global food security and nutrition. In this scientometric study, we examine current trends and identify future research directions concerning the impact of climate change on capture fisheries and fish landings. We analyzed 3164 publications from the Web of Science Core Collection database utilizing quantitative metadata analysis and scientometric techniques. Our findings reveal that the United States and Japan are the largest contributors and exert the most influence in this research area. Highly cited articles indicate that climate change has the potential to significantly redistribute global fisheries catch, with up to a 70% shift, particularly affecting high latitude regions and the tropics. Prominent and influential keywords within the knowledge base encompass "climate change-management-impact" and "climate change-small scale fishery-impact." A cluster analysis of article titles identifies the top three research foci as future trajectories, food security, and small pelagic fish. Through our scientometric investigation, we found that addressing the impacts of climate change on small-scale fisheries and maintaining food security under changing environmental conditions are among the most critical research priorities. By understanding the current research landscape and identifying emerging trends, researchers can better collaborate and address critical questions related to the effects of climate change on capture fisheries and fish landings, ultimately contributing to more sustainable and resilient fisheries management.

Keywords: small-scale fisheries; pelagic fishes; food security; seafood trade; fisheries prediction; future projection; bottom line; seasonal food

1. Introduction

Climate change has emerged as one of the most pressing challenges facing the global fishery sector, with far-reaching implications for food security, economic well-being, and the livelihoods of millions of people worldwide (FAO, 2022). The fishery sector plays an essential role in global food security, with per capita fish consumption increasing from 9.0 kg in 1961 to 20.5 kg in

2017 (FAO, 2018). However, the sustainability of these vital resources is increasingly threatened by the compounding effects of climate change, including ocean warming, acidification, and shifts in species distribution patterns (FAO, 2016). The United Nations Food and Agriculture Organization (FAO) has been monitoring global fisheries and aquaculture since 1945, providing a comprehensive dataset on the status and trends of marine resources. Despite the recent focus on sustainable fishery management, the overall decline of global fishery resources continues to be a concern (FAO, 2022). To address this challenge, the scientific community needs to better understand the complex interplay between climate change and fishery dynamics, which will ultimately contribute to the development of more effective and resilient management strategies.

The future of fishery resources is increasingly threatened by climate change projections, including ocean acidification, harmful algal blooms, and anthropogenic factors (Brander, 2010; Lam et al., 2020; Thirukanthan et al., 2023) which are anticipated to have significant repercussions on marine ecosystems and their sustainability (FAO, 2022). As such, it is crucial that scholars and policymakers continue to closely monitor, analyze, and address these pressing challenges in order to preserve and maintain the essential services provided by fisheries worldwide.

Climate change has emerged as a paramount issue confronting marine ecosystems and, consequently, posing significant challenges to capture fisheries and fish landings worldwide (Free et al., 2022). The impact of climate change on these vital resources has attracted considerable attention from the scientific community, yielding a plethora of studies exploring the complex and wide-ranging consequences. These studies have investigated various aspects, such as shifts in food chains, fish productivity, species composition, distribution, fishing mortality, and potential catches (Brander, 2007; Rijnsdorp et al., 2009; Lam et al., 2016; Heck et al., 2023). Moreover, the heterogeneity of climate change impacts across different geographic regions and fish species underscores the importance of a comprehensive and integrative analysis to effectively address these complex interactions (Cheung et al., 2010; Poloczanska et al., 2016).

Given the criticality of this subject matter and the sheer volume of research generated, there is a pressing need for a systematic, interdisciplinary approach to synthesize the existing knowledge, identify emerging trends, and guide future research directions, policy, and management decisions. Data-driven analysis, by its nature, is well-equipped to fulfill this requirement by elucidating research trends, major contributors, collaboration networks, and novel topics within the field (Chen and Song, 2019). By employing this analytical approach, this study aims to provide a comprehensive overview of the current state of research on climate change impacts on capture fisheries and fish landings, thereby offering valuable insights for policymakers, funding organizations, and stakeholders.

Several bibliometric reviews have assessed the effects of climate change on fisheries from different perspectives. For example, Xiao and Huang (2018) elucidated the influence of climate change on the fishing industry, while Shaffril et al. (2017) focused on the adaptive strategies employed by small-scale fishermen in response to evolving environmental conditions. In a recent study, Liu et al. (2021) systematically reviewed the Philippine fisheries research, revealing four main themes centered around resource management, adaptation, impact assessments, and perceptions. Their findings emphasized the need for sustainable resource management, understanding the perspectives of various stakeholders, and addressing the concerns of marginalized groups. Notably, climate change has already left a marked impact on some of the world's leading fisheries-producing species, particularly within the aquaculture sector, as evidenced by the findings of Azra et al. (2022) and Velumani et al. (2019).

Scientometric analysis offers a cutting-edge, quantitative method for examining the scientific landscape by systematically evaluating scholarly outputs, research trajectories, and the impact of academic publications. Through an assortment of metrics, such as citation analysis, network analysis, and bibliometric indicators, this innovative approach provides valuable insights into the underlying patterns, emerging trends, and collaboration networks within the scientific community (Hood & Wilson, 2001; Leydesdorff & Rafols, 2011). In contrast to traditional scientific reviews, scientometric analysis delves into the broader architecture and dynamics that define the scientific domain, thereby revealing the interconnectivity and influence of various research fields, tracking the frontiers of scientific advancement, and assessing the implications of research policies on scientific progression (Bornmann & Leydesdorff, 2014; Waltman, 2016). Leveraging the unique potential of scientometric analysis, researchers can gain a comprehensive understanding of the complex research landscape surrounding climate change, capture fisheries, and fish landing. This knowledge empowers stakeholders to develop effective adaptation strategies and promote sustainable management practices amidst ongoing environmental challenges.

In this chapter, we employ quantitative approaches and scientometric analysis to systematically investigate the existing literature on climate change and its implications for capture fisheries and fish landings. Our primary objective is to provide a comprehensive understanding of the research landscape in this field, revealing emerging trends, key contributors, and areas warranting further investigation. This knowledge will ultimately contribute to the development of effective policies and strategies for sustainable fisheries management amid a changing climate.

2. Methods

2.1. Scientometric Analysis

This chapter employs a comprehensive scientometric analysis to systematically investigate the current scenarios of climate change, capture fisheries, and fish landings. To accomplish the proposed objectives, we implement the following methodological framework:

I. **Keyword Selection:** Primary keywords and their accepted synonyms were meticulously chosen to accurately represent the terms "landing" and "captured fisheries." Multiple sources, including the Food and Agriculture Organization (FAO), were consulted for definitions and synonyms. Based on these considerations, a list of keywords and synonyms was compiled for the current study.

    Keywords for capture fisheries and fish landing:

    ("captur* fisher*") OR ("fish* land*") OR ("fish* harvest*") OR ("fish* capture*") OR ("fish* catch*") OR ("fish* caught*") OR ("fish* brought") OR ("artisanal land*") OR ("artisanal catch*") OR ("fish* trade") OR ("ocean fish*") OR ("fish* relat* activit*") OR ("catch per unit effort") OR ("fish* trawl*") OR ("seining") OR ("gillnetting") OR ("crabbing") OR ("inland fisheries") OR ("fish hunting") OR ("fish* operation") OR ("small scale fish*") OR ("indutsr* fish*") OR ("large-scale fish*") OR ("coast* fish*")

    Keywords associated with climate change:

    ("climat*") OR ("anthropogenic*") OR ("global warm*") OR ("warm* ocean*") OR ("seasonal* variat*") OR ("extrem* event*") OR ("environment* variab*") OR ("anthropogenic effect*") OR ("greenhouse effect*") OR ("sea level ris*") OR (erosio*) OR ("agricult* runoff") OR ("weather* variab*") OR ("weather* extrem*") OR ("extreme* climat*") OR ("environment* impact*")

OR (“environment* chang*”) OR (“anthropogenic stres*”) OR (“temperature ris*”) OR (“temperature effect*”) OR (“warm* ocean”) OR (“sea surface* temperat*”) OR (“heatwav*”) OR (“acidific*”) OR (“hurrican*”) OR (“el$nino”) OR (“la$ nina”) OR (“drought*”) OR (“flood*”) OR (“high precipit*”) OR (“heavy rainfall*”) OR (“CO2 concentrat*”) OR (“melt* of the glacier*”) OR (“melt* ice*”) OR (“therm* stress*”) OR (“drought”) OR (“hypoxia”) OR (“harm* alga* bloom*”) OR (“eutrophication”)

II. **Data Collection:** The Web of Science Core Collection database was queried using the “topic” search option, which searches for literature encompassing the selected keywords in the title, abstract, author keywords, and Keywords Plus. Rigorous eligibility and exclusion criteria were applied, including selecting solely research articles, excluding non-English articles, and disregarding publications from the year 2023. This process resulted in 3164 publications for qualitative analysis.

III. **Data Preparation:** The obtained data was saved as text files and prepared for subsequent scientometric analysis utilizing CiteSpace software. The data files were named “download_1_500” up to the total number of articles in the topic (i.e., 3164), resulting in a total of seven files for analysis.

IV. **Scientometric Analysis:** To conduct a comprehensive scientometric analysis of the existing literature on climate change and its implications for capture fisheries and fish landings, we employed CiteSpace software for data visualization and analysis. The software was utilized to generate various visualization maps that elucidate the structure, trends, and interconnections within the research field. The flowchart of the analysis is shown in Figure 1.

Utilizing the visualization techniques, we systematically addressed various aspects of the literature, presenting a comprehensive understanding of the research landscape:

a. **Research output:** We assessed the research output by analyzing the annual number of articles published and their citation counts. Bibliometric analyses, such as publication count and citation frequency, were used to evaluate the impact and productivity of research areas. An increasing trend in publication numbers indicates a growing interest in a specific subject, while a high citation count demonstrates the influence and impact of a study within the field.

b. **Geographical contributions:** We examined the geographical distribution of research efforts, focusing on the countries and regions involved in the research and their collaborative endeavors. Investigating the geographical distribution is essential for interpreting current findings and identifying gaps in our understanding of the subject matter.

c. **Major contributors:** We identified significant contributors in terms of funding, organizations, and disciplines to recognize potential trends and areas for further investigation. Addressing the complex nature of climate change and its effects on capture fisheries requires a collaborative approach, incorporating contributions from various disciplines, organizations, and funding agencies. Evaluating the primary contributors in this area provides crucial insights into the current research landscape and unveils potential areas for further exploration.

d. **Influential articles and authors:** We determined the influential articles and authors through citation analysis, offering insights into significant works and leading experts in the field. This analysis allows for a better understanding of the foundational research and the prominent researchers shaping the discourse on climate change impacts on capture fisheries and fish landings.

V. **Cluster Analysis:** A sophisticated cluster analysis was conducted to identify latent semantic themes in the data and to group research documents based on term correlations. The homogeneity of a cluster can be quantified using an index called the mean silhouette, which ranges from -1 to 1, with higher values indicating a more similar set of members within a cluster (Chen and Morris, 2003; Zhong et al., 2019).

VI. **Timeline Co-citation Analysis:** Timeline co-citation analysis was utilized to scrutinize the relationships between co-cited literature across specific time periods. This approach assists in identifying patterns of scientific collaboration, knowledge distribution, recent developments, and future research directions (Chen, 2022).

VII. **Keyword Distribution Analysis:** A comprehensive analysis of keyword co-occurrence in research documents was performed to identify research trends over time. This method highlights areas of research currently receiving significant attention and can reveal understudied or overlooked areas that may warrant further investigation (Radhakrishnan et al., 2017; Zou et al., 2022).

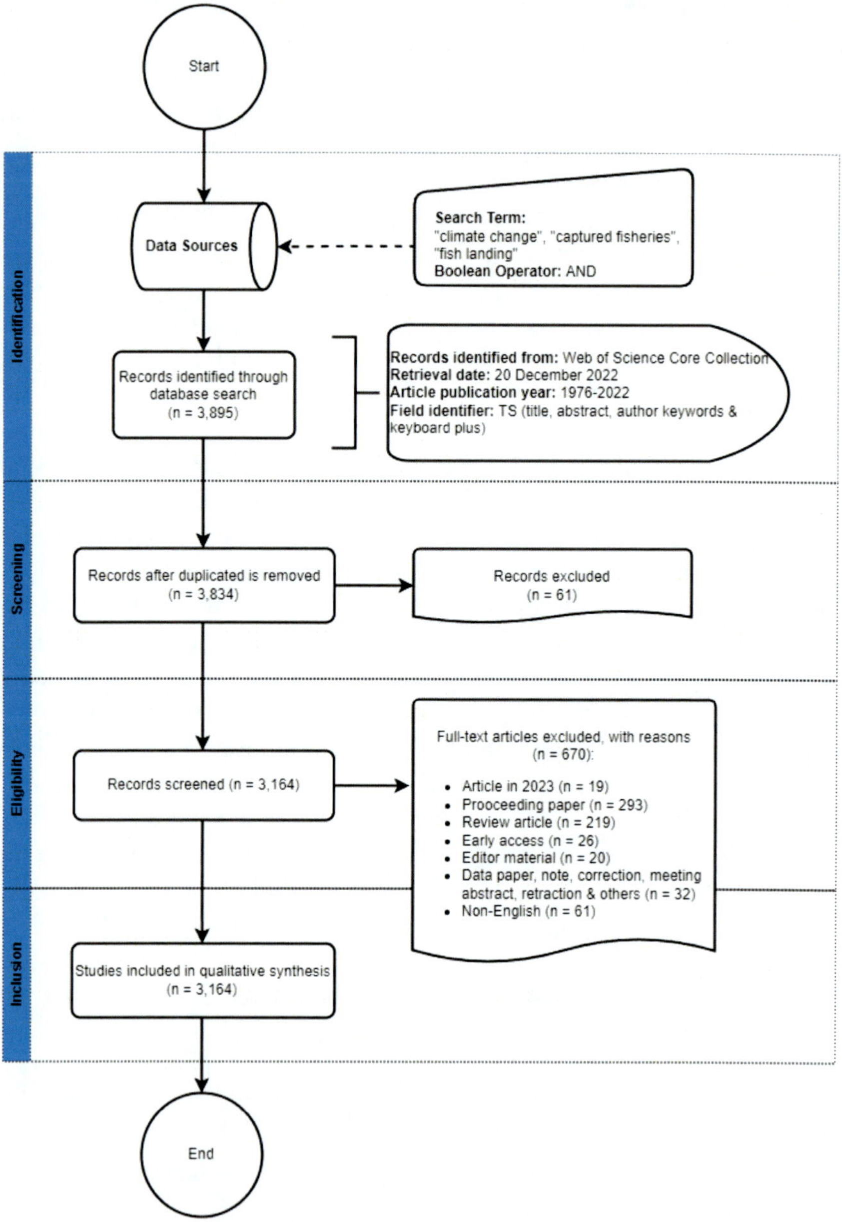

Figure 1. Flowchart for the study on captured fisheries and fish landing, associated with climate change.

3. Results

3.1. Temporal Trends in Publication Output

Our investigation revealed that the total number of articles addressing the intersection between climate change and capture fisheries and fish landing witnessed a 73% increase in 2021 compared to a decade prior (2011) (Figure 2). This observed growth indicates that the research area is rapidly expanding, drawing in a larger number of researchers and research groups, potentially due to augmented funding and resource allocation. Moreover, this suggests that the field maintains an active status with substantial engagement from the scientific community.

Furthermore, the analysis identified 131,915 references generated from the 3,164 articles, averaging 42 citations per article. These metrics offer valuable insights into the expansion and impact of the field, emphasizing the importance of considering both publication and citation counts in bibliometric-oriented studies.

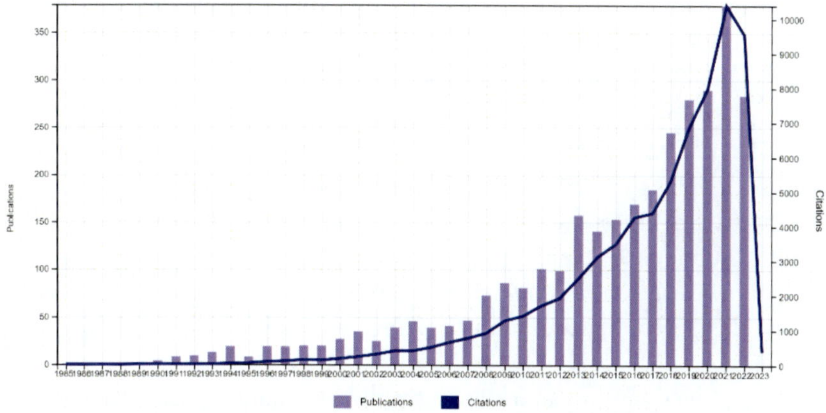

Figure 2. Number of articles and citations on the impact of climate change on capture fisheries and fish landings related research between 1985 and 2022.

3.2. Geographical Distribution of Contributions

The chapter represented 156 different regions or countries, with the United States producing the most publications overall. Canada, the United Kingdom (including England, Wales, Scotland, and Northern Ireland), Australia, and Brazil all followed closely behind (Figure 3). The participation of numerous countries highlights the significance of the field of climate change and its intersection with captured fisheries and fish landing in terms of cultural and societal context.

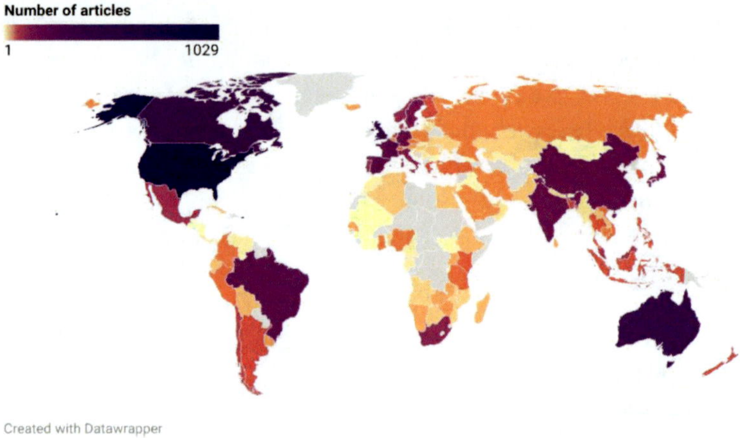

Figure 3. The countries of regions involved in the impact of climate change on capture fisheries and fish landings related research between 1985 and 2022, generated from the online map generator, app.datawrapper.de based on the Web of Science Core Collection database

The majority of studies were conducted in the Americas, whereas African regions displayed a noticeably lower research output. This disparity may be attributed to differences in research funding, infrastructure, and the prioritization of research agendas. Furthermore, our analysis revealed a scarcity of research in landlocked countries such as Niger, Chad, and Mali, where the direct impact of climate change on marine resources may not be as evident.

Additional investigations are warranted in countries with limited coastal regions, particularly those in the developing world like Libya and Somalia. The lack of research in these areas could be attributed to insufficient funding or other challenges related to research capacity. Understanding each country's engagement in researching the effects of climate change on capture fisheries and fish landing offers invaluable insights into the global research landscape. This knowledge can help direct resources and international collaboration to address underrepresented regions and advance our collective understanding of the complex interrelationships between climate change, fisheries, and fish landing.

3.3. Comprehensive Analysis of Key Contributors

Figure 4 presents a cooperation network clustering, generated by CiteSpace, illustrating the connections among researchers and institutions. Figure 5 presents the visualization map to showcase the range of disciplinary categories embodied by the 3,164 articles scrutinized in this study.

Table 1 indicates the top 10 important research disciplines in terms of the number of published articles, their affiliated organizations, and the major funding agencies. The data suggest that Marine Freshwater Biology and Fisheries are the main contributors, with 634 and 571 published articles, respectively. The National Oceanic Atmospheric Administration (NOAA) in the USA is the most published organization, followed by the University of British Columbia and the University of California. The European Commission is the leading grant provider, followed by the Natural Sciences and Engineering Research Council of Canada and the National Natural Science Foundation of China.

3.4. Prominent Authors and High-Impact Articles

Table 2 and Table 3 lists the top 10 highly cited authors and their articles in the related disciplines. The authors, hailing from various prestigious institutions, have made significant contributions to the field, furthering our understanding of the complex interplay between climate change, marine ecosystems, and human activities.

Leading the list is William Cheung from the Institute for the Oceans and Fisheries at the University of British Columbia, Canada, with 36 publications. Cheung's research has been instrumental in predicting shifts in global catch potential due to climate change and examining the relationships between mean temperature of catch (MTC) and regional sea surface temperature changes. Daniel Pauly, also from the University of British Columbia, follows with 26 publications. Pauly's work has focused on global fisheries management, sustainable exploitation, and the effects of climate change on fish stocks. His research has provided invaluable insights for policymakers and the scientific community. Xiujuan Chen from the University of Texas at Arlington, USA, ranks third with 25 publications. Chen's research interests include marine ecology, biological oceanography, and the implications of climate change on marine ecosystems.

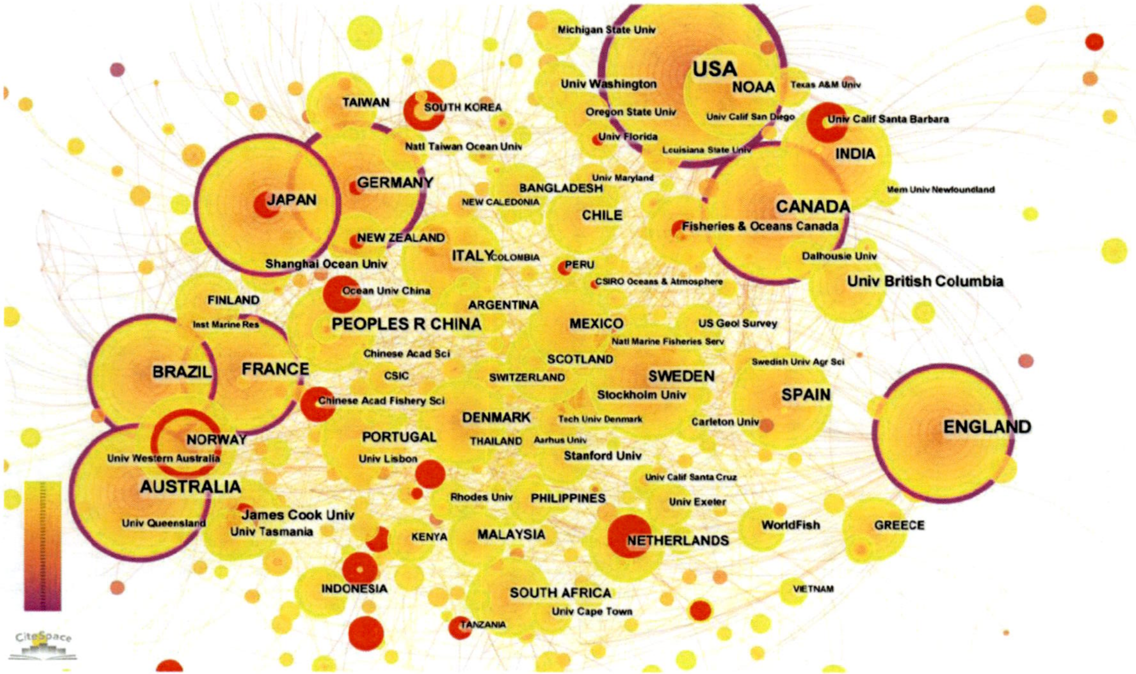

Figure 4. Collaborative network of countries and organizations in climate change, capture fisheries, and fish landing research from 1985 to 2022: Node size reflects the volume of publications in each country or region, while connection thickness indicates the degree of cooperation between them. Nodes with larger purple outer circles serve as network hubs, connecting to a high number of collaborative networks.

Table 1. Main contributor in terms of research disciplines, organization and funding agencies in the field of climate change associated with captured fisheries and fish landings

Important Research Disciplines		Most Published Organization		Major Funding Agencies	
Research Disciplines	No.	Affiliation	No.	Grant providers	No.
Marine Freshwater Biology	634	National Oceanic Atmospheric Administration, NOAA, USA	140	European Commission	86
Fisheries	571	University of British Columbia	104	Natural Sciences and Engineering Research Council of Canada	63
Environ-mental Sciences	488	University of California	73	National Natural Science Foundation of China	62
Oceano-graphy	371	Fisheries Oceans Canada	59	National Science Foundation, NSF	62
Ecology	370	United States Department Of The Interior	59	UK Research Innovation, UKRI	56
Environ-mental Studies	159	Centre National de la Recherche Scientifique (CNRS), France	55	CGIAR	54
Water Resources	144	State University System of Florida	53	Natural Environment Research Council, NERC	48
Multi-disciplinary Sciences	125	The Research Institute for Development, France	51	National Council for Scientific and Technological Development, Brazil	46
Biodiversity Con-servation	108	United States Geological Survey	48	Coordenação de Aperfeiçoamento de Pessoal de Nível Superior, CAPES, Brazil	43
Zoology	70	University of Washington	48	National Oceanic Atmospheric Administration, NOAA, USA	41

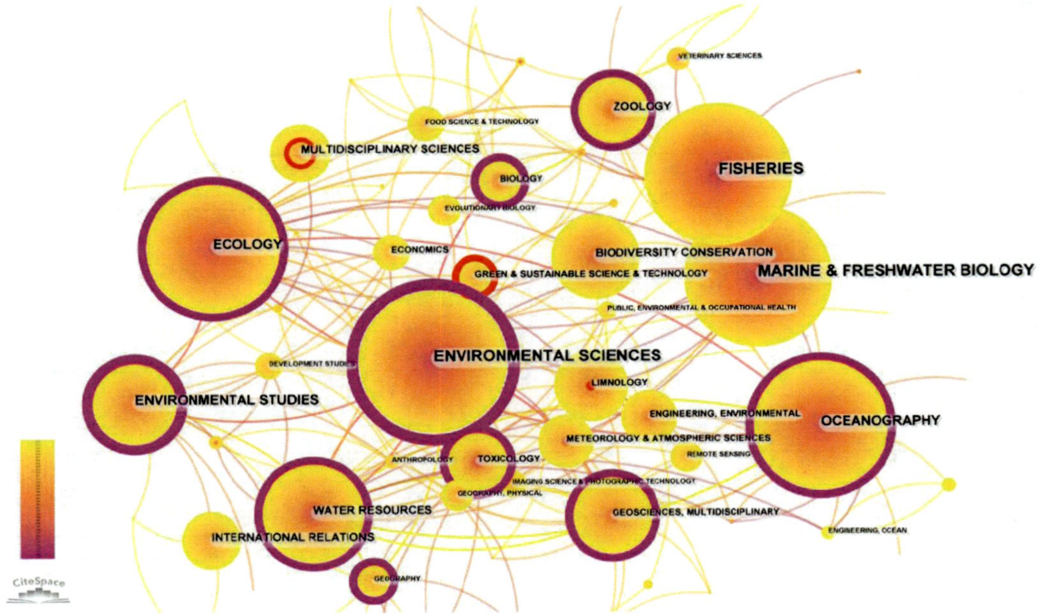

Figure 5. Interdisciplinary connections in climate change, capture fisheries, and fish landing research from 1985 to 2022: Node sizes represent the frequency of subject category co-occurrences, and line thickness signifies the strength of linkages between research disciplines. Nodes with larger purple outer circles serve as network hubs, connecting to a high number of other research disciplines.

Steven J. Cooke from Carleton University, Canada, has contributed to the field with 21 publications, focusing on the conservation and management of recreational fisheries, as well as the broader implications of climate change on fish populations. Dirk Zeller from the University of Western Australia has 16 publications, with research interests spanning the assessment of global marine fisheries resources and the development of innovative methods for estimating historical fish catch data. Vicky Wing Yee Lam and Gabriel Reygondeau, both from the University of British Columbia, Canada, have each contributed 15 publications. Lam's research covers the economic, social, and environmental aspects of global fisheries, while Reygondeau focuses on biogeography and the effects of climate change on marine ecosystems.

Ian Cowx from the University of Hull, United Kingdom, has 14 publications, concentrating on the management and conservation of inland fisheries, as well as the application of fishery science to support sustainable fisheries management. Tian Yongjun from the Ocean University of China also has 14 publications, with research exploring fish stock assessment, fisheries management, and the response of marine ecosystems to climate change. Lastly, Yingying Chen from the Chinese Academy of Science, China, has contributed 12 publications, focusing on marine biodiversity, fishery resources, and the ecological impacts of climate change on marine ecosystems.

Table 2. Top 10 highly cited authors in the field of impact of climate change on capture fisheries and fish landings related research between 1985 and 2022

No.	Author name	Affiliation	Count
1.	William Cheung	Institute for the Oceans and Fisheries, The University of British Columbia, Canada	36
2.	Daniel Pauly	University of British Columbia, Canada	26
3.	Xiujuan Chen	University of Texas at Arlington, USA	25
4.	Steven J. Cooke	Carleton University, Canada	21
5.	Dirk Zeller	University of Western Australia	16
6.	Vicky Wing Yee Lam	University of British Columbia, Canada	15
7.	Gabriel Reygondeau	University of British Columbia, Canada	15
8.	Ian Cowx	University of Hull, United Kingdom	14
9.	Tian, Yongjun	Ocean University of China	14
10.	Yingying Chen	Chinese Academy of Science, China	12

Table 3. Top 10 articles in the field based on the Web of Science Core Collection database, with the article's main conclusion in the impact of climate change on capture fisheries and fish landings related research between 1985 and 2022

No.	Reference	Journal	Total Citation	Main Conclusion
1.	Cheung et al., 2010	Global Change Biology	726	Climate change may redistribute global catch potential, with high-latitude regions gaining 30–70% and the tropics losing up to 40%.
2.	Brander, 2007	Proceedings of the National Academy of Sciences	558	Changes in climate variability amplitude are likely to have greater impacts than mean values. Extreme climate events affect marine and inland fisheries.
3.	Post et al., 2002	Fisheries	515	Fishery scientists, managers, and the public have ignored decades-long declines in four high-profile Canadian recreational fisheries.
4.	Cheung et al., 2013	Nature	498	Between 1970 and 2006, global and non-tropical MTC increased by 0.19 and 0.23 degrees Celsius per decade. MTC changes in 52 large marine ecosystems covering most of the world's coastal and shelf areas are significantly and positively related to regional sea surface temperature changes.
5.	Hutchings and Myers, 1994	Canadian Journal of Fisheries and Aquatic Sciences	425	The collapse of northern cod was solely due to overexploitation, and population sustainability indices can be used to assess exploited populations' susceptibility and resilience and reduce their commercial extinction risk.
6.	Béné et al., 2015	Food Security	378	A review of current and future fisheries and aquaculture debates and disputes, as well as a discussion of these topics in the context of agriculture and farming literature.
7.	Béné et al., 2016	World Development	376	Fish contributes to nutrition and food security, but the relationship between fisheries/aquaculture and poverty alleviation is complicated. Based on these results, this paper identifies six policy, development, and research gaps.
8.	Mason et al., 2012	Environmental Research	367	Human exposure to mercury from migratory pelagic fish like tuna and swordfish is reduced by reducing anthropogenic Hg releases and ocean deposition.
9.	Last et al., 2011	Global Ecology and Biogeography	346	Tasmanian seas have lost some of the largest predatory reef fishes since the "late 1800s" due to irresponsible fishing. In addition, Tasmanian fish distribution patterns have changed dramatically due to local marine environment warming.
10.	Breitburg, 2002	Estuaries	339	Nutrient enrichment typically increases prey abundance in surface waters with higher oxygen levels outside the hypoxic zone. Hypoxia's effects on fish, habitat, and food webs may make them more vulnerable to anthropogenic and natural stressors.

In summary, the authors have made significant contributions to the field and have provided important insights into the challenges facing marine, and freshwater biology, ecology, and oceanography. In addition, the majority of the highly cited articles discussed how anthropogenic changes to the environment have an impact on fisheries catch, fish distribution, habitat, and food webs.

This scientometric analysis highlights the top 10 impactful research articles addressing the impact of climate change on capture fisheries and fish landings between 1985 and 2022 (Table 3). The articles reveal varying perspectives, methodologies, and conclusions on the subject, with some overarching themes emerging. Firstly, the redistribution of global catch potential due to climate change is a critical concern. Cheung et al. (2010) predict that high-latitude regions could gain 30-70% in catch potential, while tropical areas may lose up to 40%. This is further supported by Cheung et al. (2013), who found that mean temperature of catch (MTC) changes in 52 large marine ecosystems are positively related to regional sea surface temperature changes. Secondly, the implications of climate variability on fisheries are significant. Brander (2007) emphasizes that changes in climate variability amplitude may have greater impacts than mean values, particularly in the context of extreme climate events affecting marine and inland fisheries. The role of overexploitation in fishery decline is also evident. Post et al. (2002) highlight the long-term decline of Canadian recreational fisheries, while Hutchings and Myers (1994) attribute the collapse of the northern cod solely to overexploitation. The latter study also introduces population sustainability indices as a useful tool for assessing exploited populations' susceptibility and resilience.

Additionally, fishery research intersects with broader discussions of food security and poverty alleviation. Béné et al. (2015) provide a comprehensive review of fisheries and aquaculture debates in relation to agriculture and farming literature. In a subsequent study, Béné et al. (2016) identify complex relationships between fisheries, aquaculture, nutrition, and poverty alleviation, resulting in the identification of six policy, development, and research gaps. Environmental factors such as mercury exposure and nutrient enrichment also play a crucial role in fisheries. Mason et al. (2012) suggest reducing anthropogenic mercury releases to decrease human exposure from migratory pelagic fish, while Breitburg (2002) explores the effects of hypoxia on fish, habitat, and food webs, highlighting their increased vulnerability to stressors. Lastly, Last et al. (2011) emphasize the importance of historical context, showing how Tasmanian seas have lost some of the largest predatory reef fishes due to irresponsible fishing since the late 1800s, with distribution patterns changing dramatically due to local marine environment warming.

3.5. Thematic Cluster Analysis of Research

The co-citation network of references was used in this study to examine the connections between the cited literature from 1980 to 2022. Table 4 and Figure 6 shows that the scientometric analysis produced 16 major clusters. Three algorithms, including the clustering labels for Log-Likelihood (LLR), Latent Semantic Index (LSI), and Mutual Information (MI), were used in the study to analyze the data. While LSI categorizes technical terms, LLR represents similar text content and topics. However, MI is a representation of the connections between references.

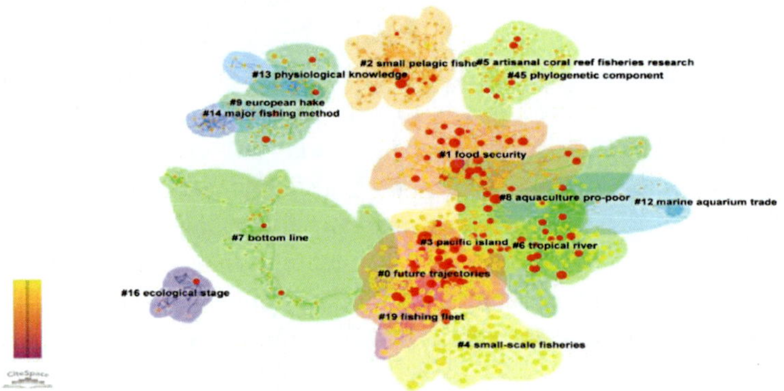

Figure 6. Reference co-citation network of climate change and fish landing-related publications: Node sizes represent citation frequency, and node colors, progressing from magenta (1985) to yellow (2022), depict the evolution of research over time. Colored connections indicate co-citation relationships, with the network further subdivided into 13 clusters through network clustering analysis.

The largest cluster (#0) has 156 members and a silhouette value of 0.881. It is labeled as "*future trajectories*" by LLR, "*small-scale fisheries*" by LSI, and "*to-demersal ratio*" *(2.67)* by MI. The major citing article of the cluster is an article that predicted the impacts of climate change and anthropogenic activities on the ocean's biophysical environment, biodiversity, and natural resources. Coll et al. (2020) introduced an updated version of EcoOcean (v2), a spatial-temporal ecosystem modeling complex of the global ocean that encompasses food web dynamics from primary producers to top predators. This system includes connections between species productivity, distribution, trophic interactions, climate change, and fishing effects. The model was utilized to simulate past and future change scenarios and compare the results using ecological indicators and biomasses of selected species groups. In their model, fishing is shown to amplify the negative effects of climate change, leading to declines in some species and mitigating the positive effects in others.

Table 4. Top 10 ranked clusters and labels produced by LSI and LLR, and MI in the impact of climate change on capture fisheries and fish landings related research between 1985 and 2022

Cluster	Size	Silhouette	Year	Label (Latent Semantic Indexing)	Label (Log Likelihood Ratio)	Label (Mutual Information Algorithm)
0	156	0.881	2017	small- scale fisheries	future trajectories	to-demersal ratio
1	142	0.883	2011	food security	food security	to-demersal ratio
2	112	0.952	2007	Atlantic multidecadal oscillation	small pelagic fish	La Niña, El Niño cycle
3	110	0.878	2015	small-scale fisheries	pacific island	adaptive coping strategies
4	64	0.939	2018	small-scale fisheries	small-scale fisheries	seafood trade
5	57	0.911	2010	future research priorities	artisanal coral reef fisheries research	fishery prediction
6	55	0.946	2016	tropical river	tropical river	severe seasonal food insecurity
7	54	0.98	2005	social cost	bottom line	marine megafauna
8	42	0.965	2013	small pelagic fish	aquaculture pro-poor	to-demersal ratio
9	36	0.98	2004	population effect	European hake	to-demersal ratio

The second largest cluster (#1) has 142 members and a silhouette value of 0.883. It is labeled as "*food security*" by both LLR and LSI, and as "*to-demersal ratio*" (1.22) by MI. This cluster is closely associated with the seventh largest cluster (#6) labeled as "*tropical river*" by both LLR and LSI, and as "*severe seasonal food insecurity" (0.51)* by MI. Fish landings, a crucial component of food security, are significantly impacted by climate change. The food security of millions of people around the world who rely on fish as their primary source of protein could be jeopardized if fish landings were to fall significantly due to changes in the ocean environment and its ecosystem (Ruckelshaus et al., 2013). Developing nations in Africa, Asia, Oceania, and Latin America appear to be most vulnerable due to their higher nutritional dependence on fish and fewer available resources to invest in climate adaptation, as reported by a global assessment (Ding et al., 2017).

The third largest cluster (#2) has 112 members, and a silhouette value of 0.952 is labeled as "*small pelagic fish*es" by LLR, "*Atlantic multidecadal oscillation"* by LSI, and "*La Niña, El Niño cycle" (0.43)* by MI. The major citing article under this cluster is by Johnson and Welch (2009), a review of the state of marine fisheries management in regard to climate change. Large-scale climate variability is known to have an effect on the populations of small pelagic clupeid fish like anchovies and sardines (Rykaczewski and Checkley, 2008). A recent study (Báez et al., 2022) examined potential connections between the landings of European anchovy and sardine and the South Oscillation Index, Pacific Decadal Oscillation, and regional climate oscillations (i.e., the Atlantic Multidecadal Oscillation, the North Atlantic Oscillation, the Western Mediterranean Oscillation index, and the Arctic Oscillation). The findings indicated that sardine landings were positively impacted by the Pacific Decadal Oscillation, whereas the Western Mediterranean Oscillation Index and the Atlantic Multidecadal Oscillation positively impacted anchovy landings. However, sardine landings, biomass, and abundance in the northwestern Mediterranean Sea have declined significantly in recent decades.

The fourth largest cluster (#3) has 110 members and a silhouette value of 0.878. It is labeled as "*pacific island"* by LLR, "*small-scale fisheries"* by LSI, and "*adaptive coping strategies" (1.76)* by MI. The major citing article under this cluster is by Frawley et al. (2019). This article examines the impact of globalization on small-scale fisheries around the world, with a focus on the Gulf of California. It examines the transformation of the social fabric of coastal fishing communities and how neoliberal reforms, including the international seafood trade, have influenced the institutional and environmental integrity of the area. According to the authors, small-scale fishermen, coastal communities, and the marine environment have all been repositioned due to the expansion of the export-oriented seafood industry. Their findings imply that tackling the tragedy of the commons in small-scale fisheries systems necessitates taking into account the interconnections between international trade, economic and social inequalities, and the potential for community-based solutions.

The fifth largest cluster (#4) has 64 members and a silhouette value of 0.939. It is labeled as "*small-scale fisheries*" by both LLR and LSI and as "*seafood trade*" *(0.3)* by MI. The major citing article under this cluster is by Salgueiro-Otero et al. (2022), looking at the role of social-ecological networks in small-scale fisheries. About 90% of fishers worldwide engage in small-scale fisheries, which provide half of the marine fish that are caught for human consumption (Thorpe et al., 2006). The impact of climate change on small-scale fisheries is a crucial topic that has been widely discussed, as these types of fishing operations are a significant source of livelihood and food security for many communities. Research has shown that changes in climate conditions can alter the abundance and distribution of fish populations, affecting the viability of small-scale fishing operations (Frawley et al., 2019). Addressing the shortcomings and limitations of small-scale fisheries, such as limited access to resources and market opportunities, is crucial for ensuring the long-term viability of these operations and the communities they support (Le Cornu, 2018).

The sixth largest cluster (#5) has 57 members and a silhouette value of 0.911. It is labeled as "*artisanal coral reef fisheries research*" by LLR, "*future research priorities*" by LSI, and "*fishery prediction*" *(0.1)* by MI. Johnson et al. (2013), the most-cited article in this cluster, conducted a systematic review of current and emerging research priorities in the field of artisanal coral reef fisheries. Artisanal fishing, which is prevalent in the Caribbean, the Americas, Asia, and Africa in particular, is the main cause of the decline in reef fish populations (Newton et al., 2007). The impacts of stressors, such as climate change, on coral reef ecosystems, both directly and indirectly, need to be better understood (Bascompte et al., 2005). Elevated water temperatures, ocean acidification, and sea-level rise brought about by climate change can significantly impact the reproductive biology and growth rates of coral reef species and may render conventional models of population and ecosystem dynamics obsolete (Hoegh-Guldberg and Bruno 2010). Additionally, the productivity and metabolic rates of species are impacted by climate change, which causes some fisheries to shift poleward (Cheung et al., 2010; Kroeker et al., 2010). In some cases, heavily fished reefs result in the dominance of rapidly growing herbivorous species, which indirectly affects food security (McClanahan, 2010).

3.6. Timeline Co-citation Analysis

In recent years, there has been a substantial increase in the publication of research examining the impact of climate change on capture fisheries and fish landings. Five distinct themes have emerged as the most prominent research areas (Figure 7), including (#0) "future trajectories", (#3) "Pacific Island", (#4) "small-scale fisheries", (#6) "tropical river", and (#8) "aquaculture". Due to their substantial impact on the industry, these subjects have become hot topics in climate change research and capture fisheries.

Projecting the long-term effects of climate change on capture fisheries and fish landings is the main goal of the "future trajectories" theme. The need to plan for the future and adapt to the changing environmental conditions has drawn a lot of research attention to this theme. In addition, the unique difficulties this region faces with regard to both climate change and fishing have drawn much attention to the theme of "Pacific Island." The impact of climate change on this industry has grown to be a major concern as Pacific Island nations rely heavily on fishing for food security and economic benefits.

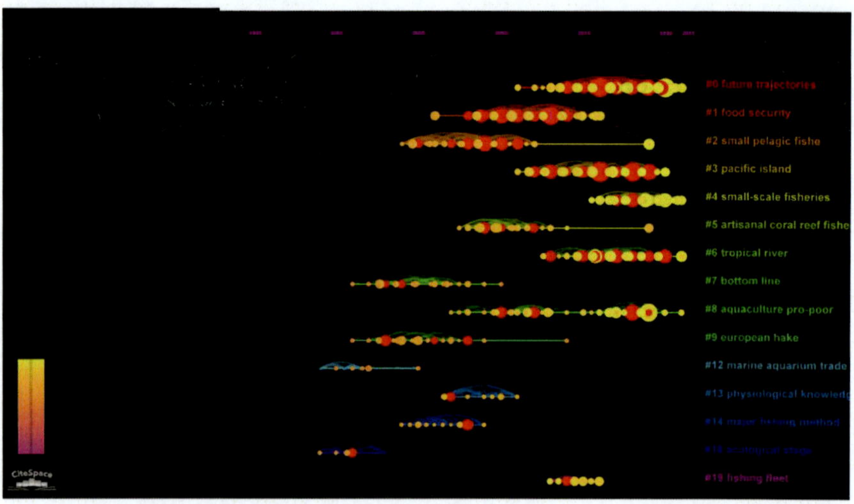

Figure 7. Timeline co-citation analysis of climate change, capture fisheries, and fish landing research from 1985 to 2022: Nodes represent reference names, and lines depict connections between them. Larger nodes signify higher citation frequencies. References with strong citation bursts are marked by red rings, while those with high centrality feature yellow rings. Longer colored line segments in the figure correspond to more extensive citation time spans.

The "small-scale fisheries" theme draws attention to the effects of climate change on small-scale fishing communities, particularly in developing nations. Due to the vital role that small-scale fisheries play in ensuring worldwide food security and aiding in the fight against poverty, this topic has taken on greater significance. The "tropical river" theme focuses on how climate change will affect the tropical freshwater fishing industry. This theme has gained more significance as riverine systems deal with various issues, such as shifting water flows, rising temperatures, and altered water chemistry, all of which significantly impact fish populations and fishing methods. "Aquaculture", as a theme, also addresses how climate change affects the development and production of farmed fish and other aquatic species. Given that aquaculture is currently the world's primary source of fish for human consumption and is growing quickly, this topic is of utmost significance. Understanding how climate change will affect aquaculture is therefore crucial.

The connections between these five themes emphasize the need to take a comprehensive approach to mitigating climate change's effects on capture fisheries and fish landings. In addition, the complex interactions between fishing practices, climate change, and the health of the world's marine and freshwater ecosystems require more study.

3.7. Keywords Distribution and Research Focus

CiteSpace was employed to analyze and visualize the recurrence and relative frequency of keywords in the capture fisheries and fish landings research from 1985 to 2022. The analysis identified 847 keywords and 3,142 links, resulting in 13 keyword clusters that represent the research hotspots and frontiers in this domain (Figure 8). These clusters facilitate a refined classification of the literature by generating a distinct set of descriptors for each cluster (Table 5).

The largest cluster (#0) has 166 members and a silhouette value of 0.644. It is labelled as *"Pacific Ocean"* by both LLR and LSI and as *"variance contribution" (2.55)* by MI. The major citing article under this cluster is by Alheit et al. (2014). The article emphasized how the small pelagic clupeoid fish populations are affected by the regime shift in the eastern North and Central Atlantic ecosystems. Fishery, abundance, growth, temperature, and ecosystem are some of the main words used to describe this cluster.

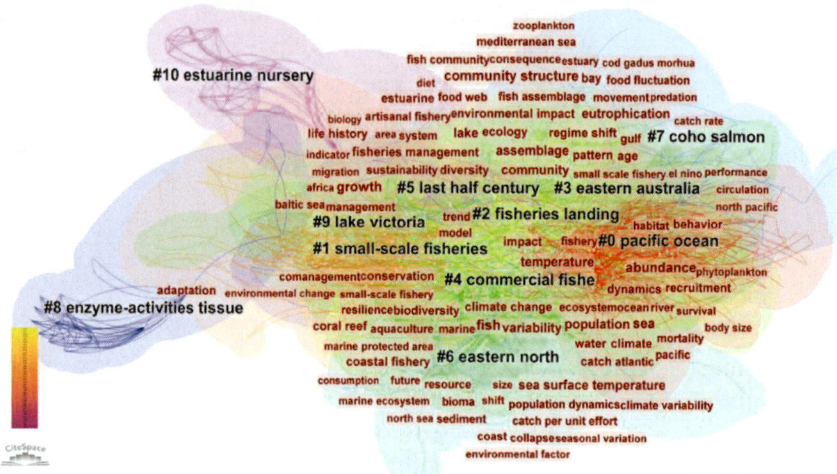

Figure 8. Distribution of co-cited clustered keywords in the impact of climate change on capture fisheries and fish landings related research between 1985 and 2022.

The second largest cluster (#1) has 131 members and a silhouette value of 0.697. It is labelled as *"small-scale fisheries"* by both LLR and LSI and as *"variance contribution" (1.9)* by MI. The major citing article under this cluster is by Galappaththi et al. (2020). The case study investigates how native Coastal-Vedda fishers in the eastern province of Sri Lanka adapt to the changing climate. They highlighted the four key components of their adaptation strategies: cultural identity, fisheries management based on culture, adaptability in selecting options, and indigenous and local knowledge systems and learning. The major keywords used to describe this cluster (#1) are management, conservation, vulnerability, food security and fisheries management.

The third largest cluster (#2) has 108 members and a silhouette value of 0.682. It is labelled as *"fisheries landing"* by LLR, *"fish assemblage"* by LSI, and *"variance contribution" (0.87)* by MI. The most-cited article in this group is Mamauag et al. (2013), which evaluated the vulnerability of tropical coastal fisheries ecosystems to climate change using the vulnerability assessment for understanding the resilience of fisheries framework (VA-TURF). The major keywords used to describe this cluster are biodiversity, assemblage, ecology, marine protected area and community structure.

The fourth largest cluster (#3) has 84 members and a silhouette value of 0.675. It is labelled as *"eastern Australia"* by LLR, *fish assemblage* by LSI, and *"variance contribution" (1.02)* by MI. This cluster is closely associated with the tenth largest cluster labelled *"lake Victoria"* by LLR, *"fish assemblage"* by LSI, and *"first offshore wind farm" (0.03)* by MI. The most important cited article in this cluster is Gelwick (1990), which investigated the fish assemblages in riffles and pools in Battle Branch, Oklahoma, and found that temporal patterns were connected to changes in species richness and fish abundance, particularly of juveniles. The major keywords used to describe this cluster (#3) are community, pattern, habitat, population, river and fish assemblage. The major keywords used to describe this cluster (#9) are freshwater, habitat use, risk, brown trout, rainbow trout and organic matter.

The fifth largest cluster (#4) has 83 members and a silhouette value of 0.699. It is labelled as *"commercial fishes"* by LLR, *"seasonal variation"* by LSI, and *"variance contribution" (0.97)* by MI. This cluster is closely associated with the sixth largest cluster (#5), labelled as *"last half century"* by LLR, *"Baltic sea"* by LSI, and *"variance contribution" (0.68)* by MI as well as the eighth largest cluster (#7), labelled as *"coho salmon"* by both LLR and LSI, and as *"size condition" (0.04)* by MI. Understanding seasonal patterns in the abundance of commercial fish species and their contribution to the overall catch is essential for conserving and maintaining sustainable fishing practices. The high MI value for "variance contribution" emphasizes the significance of examining the contribution of individual species to the overall variability in fish populations and the effect of these fluctuations on the commercial fishing industry. The major keywords used to describe this cluster (#4) are fish, marine, lake, response, food web, water quality, sediment, diet, scale, estuary and carbon. The major keywords used to describe this cluster (#5) are impact, regime shift, environmental impact and food. The major keywords used to describe this cluster (#7) are performance, Atlantic salmon, pacific salmon, exposure and bycatch

The ninth largest cluster (#8) has 23 members and a silhouette value of 0.975. It is labelled as *"enzyme-activities tissue"* by LLR, *"small-scale fisheries"* by LSI, and *empirical evidence (0.02)* by MI. The major citing article under this cluster is by Koch et al. (1992), which reviewed the metabolic responses of enzymes respond to season and temperature variations. The major keywords used to describe this cluster (#8) are adaptation, network, social capital, small-scale fisher and acclimation.

Table 5. Most frequent keyword label describing the cluster label for capture fisheries-fish landings and climate change publications (1976 – 2022)

Cluster	Cluster label	Keyword descriptors
#0	Pacific Ocean	Fishery, abundance, growth, temperature, ecosystem, dynamics, model, ocean, sea, climate, recruitment, catch, sea surface temperature, behaviour, migration, size, population dynamics, mortality, survival, catch per unit effort, El Niño, age, shift, movement, generalized additive model, bay, gulf, pacific
#1	Small scale fisheries	Management, conservation, small-scale fishery, vulnerability, food security, fisheries management, resilience, sustainability, system, aquaculture, co-management, ecosystem service, coral reef, coastal fishery, framework, coastal, inland fishery, adaptive capacity, livelihood, governance, artisanal fishery, resource, environmental change, future, indicator, protected area, marine fishery
#2	Fisheries landing	Biodiversity, assemblage, ecology, marine protected area, community structure, trend, Atlantic, estuarine, biomass, marine ecosystem, fish community, Mediterranean Sea, connectivity, species richness, collapse, environmental variable, continental shelf, South Africa, Brazil, cod, coastal fishes, marine fish, wetland, hydropower, freshwater fish, citizen science
#3	Eastern Australia	Community, pattern, habitat, population, river, fish assemblage, eutrophication, Baltic Sea, consequence, distribution, floodplain, stock, land use, predation, reproduction, Australia, sea level rise, assemblage structure, evolution, oscillation, nutrient, freshwater flow, decline, fish population, coastal fish, chlorophyll-a, flood pulse, deforestation
#4	Commercial fishes	Fish, marine, lake, response, food web, water quality, sediment, diet, scale, estuary, carbon, environment, seasonal variation, quality, environmental factor, nitrogen, coastal lagoon, stable isotope, biodiversity conservation, pollution, gear, local ecological knowledge, bioaccumulation, contamination, artisanal fishing, restoration, phosphorus, coastal management
#5	Last half-century	Impact, regime shift, environmental impact, food, cod *Gadus morhua*, North Sea, disturbance, life cycle assessment, trophic cascade, hypoxia, tuna, harmful algal bloom, extinction, salmon, long-term change, great lake, ecosystem-based management, Atlantic mackerel, brown shrimp, bird, value chain, emission, western, common, Laurentia great lake, nutrient enrichment, *Thunnus alalunga*
#7	Coho salmon	Performance, Atlantic salmon, Pacific salmon, exposure, bycatch, dissolved oxygen, identification, carbon footprint, forage fish, chinook salmon, Taiwan, *Caretta caretta*, mesh size, coho salmon, habitat degradation, efficiency, fish farm, rainfall
#8	Enzyme activities tissue	Adaptation, network, social capital, small-scale fisher, acclimation, seafood trade, gender, perspective, myofibrillar atp-ase, digestive enzyme, teleost, body mass relationship, larvae, acetoacetyl-coa thiolase. Cytochrome oxidase, aquatic pollution, *Carassius auratus*, *Scardinius erythrophthalmus*, Cyprinidae, oxygen consumption, reproductive cycle
#9	Lake Victoria	Freshwater, habitat use, risk, brown trout, rainbow trout, organic matter, black sea, fish catch, species diversity, productivity, species distribution, freshwater biodiversity, regression, yellow perch, east Africa, biomarker response, multiple stressors, *Mugil cephalus*, marine sediment

4. General Discussion

As climate change persistently alters marine ecosystems, the fishing industry faces significant repercussions. Global fish landings experience a decline due to factors such as shifting ocean currents, ocean acidification, and escalating sea surface temperatures, which consequently modify fish population distribution and abundance. To devise efficacious management strategies, understanding the current scientific knowledge concerning climate change's impact on fish populations and the subsequent implications for global fish landings is crucial. Implementing climate-adaptive fisheries management approaches, such as ecosystem-based fisheries management (EBFM) and the Resist-Accept-Direct (RAD) framework, can bolster the resilience of fisheries and marine ecosystems in the face of climate change (Pikitch et al., 2004; Plagányi, 2007).

Recent studies indicate that climate change-driven alterations in fish populations are transforming their distribution and abundance. For instance, tropical fish species extend their range poleward in response to increasing ocean temperatures, while cold-water species contract their range (Benoit et al., 2023). These range shifts present challenges for fishery management as traditional fishing grounds become less productive and significantly affect fish landings.

Moreover, ocean acidification, resulting from the absorption of atmospheric carbon dioxide by the ocean, poses a threat to fish populations. As seawater acidity increases, fish populations decline, creating a hostile environment for marine life. Ocean acidification disrupts the availability of specific plankton species, vital food sources for fish, thus inducing ripple effects throughout the entire food web (Cooke et al., 2022). Adopting the RAD framework in fisheries management, which involves resisting change, accepting change, or directing change, can help stakeholders adapt to these dynamic conditions and mitigate potential impacts (Plagányi, 2007).

In addition to inundating fishing communities, damaging boats and equipment, and altering critical coastal ecosystems for fish growth and reproduction, sea level rise impacts the fish landing industry. The increasing frequency and intensity of typhoons and hurricanes further jeopardize coastal infrastructure, complicating the process for fishermen to land their catch (Mondal et al., 2022). Consequently, incorporating climate-adaptive management strategies in the fishing industry becomes increasingly important for ensuring the sustainability and resilience of fisheries and fishing communities.

Hence, an exhaustive assessment of the scientific landscape surrounding climate change's influence on fish populations and the global fish landing consequences is paramount. Through our scientometric analysis, we aim to elucidate the core issues concerning climate change's effect on fish landings, encompassing an in-depth exploration of the underlying mechanisms and their impacts on global fish landings.

4.2. Assessing the Economic and Food Security Consequences of Climate Change on Capture Fisheries and Fish Landings

The global population relies heavily on the fishing and aquaculture industries for its food and economic stability. Fish production increased 80-fold between 1950 and 2006 and reached 214 million tonnes in 2020 (FAO, 2022). Production from capture fisheries has leveled off at about 90 million tonnes over the past decade, while production from aquaculture has increased dramatically and now makes up more than a third of total fish production (FAO, 2018). The impact of climate change on capture fisheries and fish landings is a major concern, as it has the potential to cause significant harm to food and economic security, particularly in coastal and island communities where fisheries are a primary source of livelihood and nutrition (Rice and Garcia, 2011: Techera, 2018).

According to estimates from the Food and Agriculture Organization (FAO), capture fisheries provide nearly 20% of the world's total protein for human consumption and 17% for two-fifths of the world's population. By 2050, the world's population is expected to reach 9 billion, and food security will become a significant issue on a national and international level (Béné et al., 2015). As a major source of protein for the world's poor, capture fisheries are indispensable to those living in rural areas with low incomes. More than 64 countries rely heavily on marine fisheries for more than 50% of their total production. African nations are especially vulnerable due to their high reliance on fisheries and limited ability to adapt to the effects of climate change (Ding et al., 2017). It has been found that in some coastal nations of West Africa, fish accounts for over 60% of the population's dietary protein, including 62% in Gambia and 63% in both Ghana and Sierra Leone (Béné and Heck, 2005). By the 2050s, it is expected that the annual landings of almost all West African countries will have decreased, with the exception of Gambia, Western Sahara, Mauritania, and Senegal (Lam et al., 2012).

Bangladesh, Cambodia, Maldives, Philippines, Vietnam, Thailand, and Indonesia are among the numerous Asian countries categorized as "high vulnerability" according to the vulnerability index assessment by the Intergovernmental Panel on Climate Change (McCarthy et al., 2001; Ding et al., 2017). The Philippines, in particular, has experienced a consistent decline in fish landings due to climate change impacts on key marine species such as anchovy, Indian mackerel, and tuna (Suh and Pomeroy, 2020; FAO, 2016). Studies conducted by Macusi et al. (2020, 2021) highlighted that small-scale fishers in various coastal communities, in Philippines, have already perceived the effects of climate change on their livelihoods and local ecosystems. These findings emphasize the critical role of small-scale fisheries in food security and income for coastal communities and the need for comprehensive conservation strategies. Both studies call for guidance from relevant agencies and local government units in formulating policies to increase community resilience against climate change impacts, such as unpredictable weather, coral bleaching, and changes in spawning seasons. These observations, based on perception surveys, align with the broader understanding of climate change's detrimental effects on marine ecosystems, emphasizing the urgent need for action to protect fisheries and the communities that rely on them.

The fishing industry in India is vital to the country's economy and the livelihoods of over 14.5 million people. Indian fishing communities are some of the poorest in the world despite the industry's enormous economic impact. This is due to a number of factors, including the widespread decline in fish stocks in coastal areas and market supply chains (FAO, 2016). According to a study by Mohanty et al. (2017), the effects of climate change on marine and inland fisheries and aquaculture in India will reduce ecosystem services along India's eastern coast by 25% in 25 years, leading to a cumulative loss of US$17 billion.

The Coral Triangle (CTI) region of the Asia-Pacific is home to 76% of the world's coral species and 37% of its reef fish species, making it the epicenter of marine biodiversity worldwide. Due to the 130 million people who depend on fisheries ecosystems for food, income, and livelihoods, as well as the fact that capture fisheries and aquaculture contribute 11.3% (19.1 million tonnes) of the world's fisheries, the CTI has an urgent concern for food security (Delgado, 2003; Foale et al., 2013; Cruz-Trinidad et al., 2014). Total CTI fish catches have been on the rise since 1951. Still, several studies have warned that CTI nations may have reached or even surpassed the critical carrying capacity of their demersal and pelagic fishery resources (Lymer et al., 2010). The effects of climate change in the Coral Triangle have been the subject of numerous reports. Over the past two decades, coral bleaching events in the Philippines, Indonesia, and Malaysia have reduced coral reef productivity due to rising sea temperatures and temperature anomalies (Cruz-Trinidad et al., 2014). Additionally, warming decreased primary productivity, causing an impact on small pelagic fisheries, a crucial component of the Coral Triangle's food security (Cruz-Trinidad et al., 2014; Abesamis et al., 2017)

Food security, as well as the economy and way of life of communities that depend on the marine capture fisheries sector, are significantly impacted by both the direct and indirect effects of climate change. Therefore, a major concern for food and economic security, particularly in coastal and island communities, is the effect of climate change on capture fisheries and fish landings. Further research is needed to understand the impacts of climate change on capture fisheries and to develop strategies to mitigate these impacts and ensure food and economic security for coastal and island communities.

4.3. Climate-Adaptive Fisheries Management

Climate change has numerous effects on marine ecosystems and resources. As per projections made by Lam et al. (2016), the global fisheries sector is at risk of experiencing a loss of up to $10 billion annually due to the effects of climate change. However, this scenario can be altered through effective management strategies aimed at tackling the current mismanagement issues and challenges posed by climate change. According to Gaines et al. (2018), if these reforms are implemented promptly, global fisheries can not only survive but also prosper in the future.

The development of tools for comprehending the regional impacts of climate change and the adaptive responses of marine social-ecological systems has not kept pace with that of their terrestrial counterparts. This could lead to conflicting objectives, duplicated efforts, and maladaptation, especially if the focus is on short-term gains (Allison, 2015; Holsman et al., 2019). Recent studies have started to develop strategic frameworks that list the crucial elements of a climate-ready strategy for marine systems (Busch et al., 2016).

Robust and accurate data on the state of the stocks are necessary to manage marine resources effectively. This data is crucial for assessing the current state of the resources and making well-informed decisions about how best to exploit them. The status of the resource can be determined through comparison with baselines, and this comparison is made possible through stock assessment (Jaya et al., 2022). The reliability of stock assessment results is essential for effective management, especially in the face of climate change, which can have complex effects on the population dynamics of marine species. The accuracy of the results of stock assessments can be impacted by the time-varying nature of population processes, such as migration patterns and growth rates, which can introduce retrospective biases into biomass and fishing mortality estimates (Mangano et al., 2020). To address these challenges, Schirripa et al. (2009) and Martell and Stewart (2013) proposed allowing population processes to vary in stock assessments. Szuwalski and Hollowed (2016) also emphasized the importance of performing simulation studies to better understand the impacts of mis-specifying time-varying processes in stock assessments. Three of the main approaches include Ecosystem-based Fisheries Management (EBFM), Territorial Use Rights in Fisheries (TURF), and Adaptive Fisheries Management/Active Adaptive Fisheries Management (Schirripa et al, 2009).

Ecosystem-based Fisheries Management (EBFM) considers the direct and indirect impacts of harvesting on target and non-target species, habitats, and ecological communities. EBFM promotes broader socio-economic climate change adaptation by taking into consideration social, economic, and ecological factors. It aims to enhance the resilience of natural ecosystems and ecosystem services by managing for the long term, accounting for gradual or abrupt changes (Hillborn, 2011; Costello, 2016; Busch et al., 2016; Holsman et al., 2019). Take NOAA as an example; they've adopted a policy statement and road map to implement ecosystem-based fisheries management (EBFM). The EBFM policy of NOAA encourages addressing the interconnections between ecosystems to support the maintenance of robust and productive ecosystems, even as they adapt to climatic, habitat, ecological, social, and economic changes (Townsend et al., 2019). Territorial Use Rights in Fisheries (TURF) grants access privileges and fishing rights to exploit fisheries resources within a designated area, often managed jointly by stakeholders and fisheries agencies. This right-based approach minimizes human stressors, leading to high abundance, species richness, and biomass of target and non-target species, which improves ecosystem resilience in the face of climate change (Qunyh et al., 2017). Adaptive Fisheries Management and Active Adaptive Fisheries Management prioritize iterative learning processes within decision-making to enhance the adaptive capacity of fisheries management measures in the short and long term. They seek to integrate ecological, social, and economic factors to facilitate a more comprehensive socio-economic adaptation to climate change (Edmondson and Fanning, 2022).

In a study published in 2017, Pons et al. found that high seas stocks have increased in biomass and decreased in fishing mortality over the past decade where appropriate management regimes are in place. Countries like Peru, Mexico, and Chile have improved their fisheries management by implementing commercial and subsistence fishing reforms. Future climate change is a major concern, so it is important to develop management strategies that can withstand these shifts (Kasperski et al., 2021). The Future Seas project is another emerging management modeling tool that aims to improve climate-resilient management by integrating ocean models with ecological and social models (Pozo Buil et al., 2021).

Another example is the Resist-Accept-Direct (RAD) framework, which Schuurman et al. (2022) describe as a tool for managing the uncertainty and variability associated with changing aquatic ecosystems. The framework offers three management pathways: resist the trajectory of change, accept the trajectory, or direct the trajectory. The RAD framework is useful when making management decisions because it takes into account all relevant factors across multiple scales (temporal, spatial, change magnitude, ecological, social, and financial). In Alaska, where the climate is warming, and hydrology is shifting rapidly, the RAD framework is being used to manage salmon-bearing ecosystems. Commercial salmon harvests have given way to recreational and subsistence harvests, as well as harvests for urban development. In order to effectively manage Pacific salmon in a warming world, RAD identifies the necessary ecological and social actions at various spatial scales (Lynch et al., 2022).

Conclusion and Recommendation

Employing scientometric techniques, this study offers a comprehensive examination of the existing literature concerning the global status of capture fisheries and fish landings in relation to climate change. The knowledge map generated illustrates the progression of publications, prominent contributors, research disciplines, and emerging trends and hotspots within this crucial area.

Research cluster analysis emphasizes the significance of forecasting future climate change effects on capture fisheries and fish landings while focusing on issues related to circular economy and food security. Additionally, our results suggest an increasing interest in small-scale fisheries and management research within the context of climate change.

The keyword analysis demonstrates that climate change elements, such as sea surface temperature and ocean warming, rank among the primary factors influencing global capture fisheries and fish landings, especially impacting small-scale fisheries. As a result, promoting interdisciplinary collaboration between economics and marine sciences is vital. For example, investigating the repercussions of climate change-driven coastal erosion on the income of fishermen households, particularly due to rising sea levels, presents a compelling area for further research.

Moving forward, it is essential to foster greater involvement from institutions and regions across the globe to expand scientific understanding in this field. This collaborative effort will contribute to the development of sustainable management and adaptation strategies for the fisheries sector in response to climate change challenges.

Conflict of Interest

The authors declare that the research was conducted in the absence of any commercial or financial relationships that could be construed as a potential conflict of interest.

4.1. Changes in Fish Distribution and Abundance Due to Climate-Related Factors

As climate change continues to transform global ecosystems, the scientific community has become increasingly invested in understanding its implications for capture fisheries and fish yields. The rise in water temperature associated with climate change is notably affecting the distribution and abundance of various fish species (Monnet et al., 2022; Levine et al., 2023; Benoit et al., 2023). A recent investigation (Barbarossa et al., 2021) predicts that, under a 3.2°C warming scenario, over a third of freshwater fish species will be threatened, with tropical watersheds such as the Amazon, Niger, Parana, Senegal, and Danube being particularly vulnerable. Fodrie et al. (2010) observed shifts in fish populations in seagrass meadows in the northern Gulf of Mexico between the 1970s and 2006-2007, noting the emergence of several previously absent tropical and subtropical species. This phenomenon was attributed to the poleward migration of warm-water fish in response to ocean warming. Although these short-term range shifts might initially enhance local community diversity, their long-term persistence could disrupt foraging rates of larger predators and competition for refuge or foraging spaces (Sax et al., 2007), consequently impacting capture fishery industries.

Forecasted range shifts of 45-60 km per decade indicate that 80% of fish species are moving poleward, with expected average distribution shifts of 600 km for pelagic and 220 km for demersal species by 2050 (Cheung et al., 2009). The global mean extinction rate is estimated at 0.03, with the Arctic and Southern Ocean facing the highest rates and the equatorial region experiencing the lowest. Under the high-emissions IPCC A1B scenario, high-latitude regions could see a 30-70% average increase in catch potential, while tropical regions may face decreases of up to 40% (Cheung et al., 2010). The Indo-Pacific region is anticipated to be heavily impacted, with potential declines in ten-year average maximum fisheries catch potential reaching 50% by 2055.Temperate climate species, such as salmonids, may experience negative impacts on reproductive success and population dynamics due to the loss of essential cold-tempering periods for effective spawning (Deepananda and Macusi, 2002). Moreover, increasing global temperatures are compromising fish immunity, heightening their vulnerability to diseases. Stressors like high temperatures, osmotic stress, and crowding can adversely affect fish immune function (Deepananda and Macusi, 2012). Consequently, bacterial diseases in aquaculture systems often peak at elevated temperatures (Wedemeyer and Wedemeyer, 1996). Migrating fish from warmer regions may also introduce novel parasites and diseases to native fish populations, exacerbating the consequences of climate change (Font, 2003).

Ocean acidification and oxygen minimum zone expansion threaten marine organisms and fisheries, with fish species exhibiting variable responses to acidification (Cheung et al., 2011; Pörtner and Knust, 2007; Melzner et al., 2009). Climate change-induced shifts in temperature and ocean chemistry directly affect fish growth, reproduction, and overall physiology, with more pronounced physiological changes expected in temperate and polar regions (Bailey et al., 2022; Cooke et al., 2022). Smaller maximum body sizes, earlier maturation, and higher natural mortality rates may result from warmer waters (Berggren et al., 2022). Furthermore, oxygen minimum zone expansion and ocean acidification pose risks to marine organism survival and distribution, particularly for pelagic species (Tommasi et al., 2010).

Rising sea levels threaten estuarine habitats, impacting species abundance (Hlohowskyj et al., 1996; Ehsan et al., 2019). Coastal and estuarine geomorphology changes will reduce intertidal habitat extents, diminishing benthic intertidal invertebrate populations, and ultimately affecting larger creatures like fish and epibenthic crustaceans (Crooks, 2004; Frihy and El-Sayed, 2013; Fuiji, 2012). Coastal regions like Nigeria's Lagos to Calabar South coastline face seawater intrusion, negatively impacting inland fisheries and aquaculture and destroying local livelihoods (Freeman, 2017). Coastal habitat and fishery threats due to sea-level rise are compounded by increased extreme coastal events, such as tropical cyclones and rainfall, as reported in Europe, Africa, and South Asia (Ibe and Awosika, 1991; Calafat et al., 2022; Vousdoukas et al., 2022; Mondal et al., 2022). Increased average wind speeds and wind extremes may elevate waves and intertidal habitat energy, resulting in more frequent surges and coastal flooding (Fuiji, 2012; Mondal et al., 2022). The deterioration of coastal defense infrastructure will raise maintenance costs for fishing jetties, tide gates, ports, and fishing piers, negatively impacting coastal fisheries' economic viability (Ehsan et al., 2019).

In summary, the impacts of climate change on capture fisheries and fish landings are complex and extensive, encompassing alterations in water temperature, distribution and abundance of fish species, susceptibility to diseases, and ocean chemistry.

Author Contributions

"TCS and MNA contributed to the conception and design of the study. FL organized the database. YW performed the statistical analysis. TCS wrote the first draft of the manuscript. MNA, FL and YW wrote the second revision of the manuscript. MNIK edited the manuscript according to the template. All authors contributed to manuscript revision, read, and approved the submitted version."

Funding

The study is supported by the Malaysian Ministry of Higher Education under the Long-Term Research Grant Scheme (LRGS/1/2020/UMT/01/1; LRGS UMT Vot No. 56040) entitled 'Ocean climate change: potential risk, impact and adaptation towards marine and coastal ecosystem services in Malaysia' between Universiti Malaysia Terengganu (UMT), Universiti Malaya (UM), Universiti Sains Malaysia (USM) and International Islamic University

Malaysia (IIUM). Research was supported in part by funds provided by USDA-NIFA Sustainable Agriculture Systems, Grant No. 2019-69012-29905. Title of Project: Empowering US Broiler Production for Transformation and Sustainability USDA-NIFA (Sustainable Agriculture Systems): No. 2019-69012-29905.

Acknowledgments

First author thanks the Sustainable Ocean Alliance (SOA) and Environmental Defense Fund (EDF) in the United State of America (USA) for his "Inaugural Fellow" on Leadership for Climate Resilient Fisheries (LCRF). Thanks also goes to National Research and Innovation Agency (BRIN), Indonesia for providing research and publication support through the "Visiting Researcher" for the first author, at Research Center for Marine and Land Bioindustry, West Nusa Tenggara.

Contribution to the Field Statement

Fisheries play an important role in providing food and income for coastal communities. However, recent climate change has impacted the captured fisheries and fish landing globally. This directly and indirectly causes a reduction in marine biodiversity, resources and other ecosystem services. The bibliometric based analysis here, describes the present status and current trends of available literature about the climate change associated with the captured fisheries and fish landing in the world. This review focused on the qualitative assessment and scientometric analysis of published articles in the Web of Science Core Collection database. This will help funders, scientific communities and fishermen to identify possible research direction and limited resources in the future.

References

Abesamis, R. A., Saenz-Agudelo, P., Berumen, M. L., Bode, M., Jadloc, C. R. L., Solera, L. A., ... & Russ, G. R. (2017). Reef-fish larval dispersal patterns validate no-take marine reserve network connectivity that links human communities. *Coral Reefs*, *36*, 791-801. doi: 10.1007/s00338-017-1570-0.

Alheit, J., Licandro, P., Coombs, S., Garcia, A., Giráldez, A., Santamaría, M. T. G., ... & Tsikliras, A. C. (2014). Reprint of "Atlantic Multidecadal Oscillation (AMO) modulates dynamics of small pelagic fishes and ecosystem regime shifts in the eastern North and Central Atlantic". *Journal of Marine Systems*, *133*, 88-102. doi: 10.1016/j.jmarsys.2014.02.005.

Allison, E. H., & Bassett, H. R. (2015). Climate change in the oceans: human impacts and responses. *Science*, *350*(6262), 778-782. doi: 10.1126/science.aac872.

Aranda, M. (2009). Developments on fisheries management in Peru: The new individual vessel quota system for the anchoveta fishery. *Fisheries Research*, *96*(2-3), 308-312. doi:10.1016/j.fishres.2008.11.004.

Azra M. N., Noor M. I. M., Sung Y. Y., Ghaffar MA. (2022). Visualizing a Field of Research With Scientometrics: Climate Change Associated With Major Aquatic Species Production in the World. *Frontier in Environmental Science.* 10: 905428. doi:10.3389/fenvs.2022.905428.

Bailey, L. A., Childs, A. R., James, N. C., Winkler, A., & Potts, W. M. (2022). Links between behaviour and metabolic physiology in fishes in the Anthropocene. *Reviews in Fish Biology and Fisheries*, *32*(2), 555-579. doi:10.1007/s11160-022-09701-2.

Barbarossa, V., Bosmans, J., Wanders, N., King, H., Bierkens, M. F., Huijbregts, M. A., & Schipper, A. M. (2021). Threats of global warming to the world's freshwater fishes. *Nature communications*, *12*(1), 1701. doi: 10.1038/s41467-021-21655-w.

Bascompte, J., Melian, C. J. and Sala, E. (2005) Interaction strength combinations and the overfishing of a marine food web. *Proceedings of the National Academy of Sciences of the United States of America* 102, 5443–5447. doi:10.1073/pnas.0501562102.

Báez, J. C., Pennino, M. G., Czerwinski, I. A., Coll, M., Bellido, J. M., Sánchez-Laulhé, J. M., ... & García-Soto, C. (2022). Long term oscillations of Mediterranean sardine and anchovy explained by the combined effect of multiple regional and global climatic indices. *Regional Studies in Marine Science*, *56*, 102709. doi:10.1016/j.rsma.2022.102709.

Béné, C., & Heck, S. (2005). Fish and food security in Africa. *NAGA, WorldFish Center Quarterly*, Vol. 28 No. 3 & 4 Jul-Dec 2005.

Béné, C., Barange, M., Subasinghe, R., Pinstrup-Andersen, P., Merino, G., Hemre, G. I., & Williams, M. (2015). Feeding 9 billion by 2050–Putting fish back on the menu. *Food Security*, *7*, 261-274. doi:10.1007/s12571-015-0427-z.

Béné, C., Arthur, R., Norbury, H., Allison, E. H., Beveridge, M., Bush, S., ... & Williams, M. (2016). Contribution of fisheries and aquaculture to food security and poverty reduction: assessing the current evidence. *World Development*, *79*, 177-196. doi:10.1016/j.worlddev.2015.11.007.

Benoit, D. M., Chu, C., Giacomini, H. C., & Jackson, D. A. (2023). Depth and temperature drive patterns of spatial overlap among fish thermal guilds in lakes across Ontario, Canada. *Diversity and Distributions*, *29*(2), 289-299. doi:10.1111/ddi.13661.

Berggren, T., Bergström, U., Sundblad, G., & Östman, Ö. (2022). Warmer water increases early body growth of northern pike (*Esox lucius*), but mortality has larger impact on decreasing body sizes. *Canadian Journal of Fisheries and Aquatic Sciences*, *79*(5), 771-781. doi:10.1139/cjfas-2020-0386.

Bornmann, L., & Leydesdorff, L. (2014). Scientometrics in a changing research landscape: bibliometrics has become an integral part of research quality evaluation and has been changing the practice of research. *EMBO reports*, 15(12), 1228-1232. DOI: 10.15252/embr.201439608.

Brander, K. (2010). Impacts of climate change on fisheries. *Journal of Marine Systems*, *79*(3-4), 389-402.

Brander K. M. (2007). Global fish production and climate change. *Proceedings of the National Academy of Sciences*. 104: 19709-19714. doi:10.1073/pnas.0702059104.

Breitburg, D. (2002). Effects of hypoxia, and the balance between hypoxia and enrichment, on coastal fishes and fisheries. *Estuaries*, *25*(4), 767-781. doi:10.1007/BF02804904.

Busch, D. S., Griffis, R., Link, J., Abrams, K., Baker, J., Brainard, R. E., ... & Merrick, R. (2016). Climate science strategy of the US national marine fisheries service. *Marine Policy*, *74*, 58-67. doi:10.1016/j.marpol.2016.09.001.

Calafat, F. M., Wahl, T., Tadesse, M. G., & Sparrow, S. N. (2022). Trends in Europe storm surge extremes match the rate of sea-level rise. *Nature*, *603*(7903), 841-845. doi:10.1038/s41586-022-04426-5.

Carter, R. W. G. (2013). Coastal environments: an introduction to the physical, ecological, and cultural systems of coastlines. *Elsevier*. ISBN: 9780080502144.

Chen, C., & Morris, S. (2003, October). Visualizing evolving networks: Minimum spanning trees versus pathfinder networks. In *IEEE symposium on information visualization 2003 (IEEE Cat. No. 03TH8714)* (pp. 67-74). IEEE. doi: 10.1109/ INFVIS.2003.1249010.

Chen, C., & Song, M. (2019). Visualizing a field of research: A methodology of systematic scientometric reviews. *PloS one*, *14*(10), e0223994. doi:10.1371/ journal.pone.0223994.

Chen, B., Shin, S., Wu, M., & Liu, Z. (2022). Visualizing the Knowledge Domain in Health Education: A Scientometric Analysis Based on CiteSpace. *International Journal of Environmental Research and Public Health*, *19*(11), 6440. doi:10.3390/ ijerph19116440.

Cheung, W. W., Lam, V. W., Sarmiento, J. L., Kearney, K., Watson, R., & Pauly, D. (2009). Projecting global marine biodiversity impacts under climate change scenarios. *Fish and fisheries*, *10*(3), 235-251. doi: 10.1111/j.1467-2979.2008. 00315.x.

Cheung, W. W., Lam, V. W., Sarmiento, J. L., Kearney, K., Watson, R. E. G., Zeller, D., & Pauly, D. (2010). Large-scale redistribution of maximum fisheries catch potential in the global ocean under climate change. *Global Change Biology*, *16*(1), 24-35. doi:v10.1111/j.1365-2486.2009.01995.x.

Cheung, W. W., Dunne, J., Sarmiento, J. L., & Pauly, D. (2011). Integrating ecophysiology and plankton dynamics into projected maximum fisheries catch potential under climate change in the Northeast Atlantic. *ICES Journal of Marine Science*, *68*(6), 1008-1018. doi: 10.1093/icesjms/fsr012.

Cheung, W. W., Watson, R., & Pauly, D. (2013). Signature of ocean warming in global fisheries catch. *Nature*, *497*(7449), 365-368. doi:10.1038/nature12156.

Costello, C., Ovando, D., Clavelle, T., Strauss, C. K., Hilborn, R., Melnychuk, M. C., ... & Leland, A. (2016). Global fishery prospects under contrasting management regimes.

Proceedings of the national academy of sciences, 113(18), 5125-5129. doi:10.1073/pnas.1520420113.

Crooks, S. (2004). The effect of sea-level rise on coastal geomorphology. *Ibis, 146*, 18-20. doi:10.1111/j.1474-919X.2004.00323.x.

Cruz-Trinidad, A., Aliño, P. M., Geronimo, R. C., & Cabral, R. B. (2014). Linking food security with coral reefs and fisheries in the coral triangle. *Coastal Management, 42*(2), 160-182. doi: 10.1080/08920753.2014.877761.

Coll, M., Steenbeek, J., Pennino, M. G., Buszowski, J., Kaschner, K., Lotze, H. K., ... & Christensen, V. (2020). Advancing global ecological modeling capabilities to simulate future trajectories of change in marine ecosystems. *Frontiers in Marine Science, 7*, 567877. doi: 10.3389/fmars.2020.567877.

Cooke, S. J., Auld, H. L., Birnie-Gauvin, K., Elvidge, C. K., Piczak, M. L., Twardek, W. M., ... & Muir, A. M. (2022). On the relevance of animal behavior to the management and conservation of fishes and fisheries. *Environmental Biology of Fishes*, 1-26. doi:10.1007/s10641-022-01255-3.

Deepananda, K. A., & Macusi, E. D. (2012). The Changing Climate And Its Implications To Capture Fisheries: A Review. *Journal of Nature Studies, 11*(1&2), 71-87. ISSN: 1655-3179.

Delgado, C. L. (2003). *Fish to 2020: Supply and demand in changing global markets* (Vol. 62). WorldFish. ISBN 0-89629-725-X.

Ding, Q., Chen, X., Hilborn, R., & Chen, Y. (2017). Vulnerability to impacts of climate change on marine fisheries and food security. *Marine Policy, 83*, 55-61. doi:10.1016/j.marpol.2017.05.011.

Edmondson, E., & Fanning, L. (2022). Implementing Adaptive management within a Fisheries Management Context: A Systematic Literature Review Revealing Gaps, Challenges, and Ways Forward. *Sustainability, 14*(12), 7249. doi: 10.3390/su14127249.

Ehsan, S., Begum, R. A., Nor, N. G. M., & Maulud, K. N. A. (2019). Current and potential impacts of sea level rise in the coastal areas of Malaysia. In *IOP Conference Series: Earth and Environmental Science.* 228: 012023. IOP Publishing. doi: 10.1088/1755-1315/228/1/012023.

FAO (2016). *The State of World Fisheries and Aquaculture 2016. Contributing to Food Security and Nutrition for All.* Rome: FAO. 78-89pp. ISBN: 978-92-5-109185-2.

FAO (2018). *The State of World Fisheries and Aquaculture* 2018 - Meeting the sustainable development goals. Rome. 127-227 pp. ISBN: 978-92-5-130562-1.

FAO (2022). The State of World Fisheries and Aquaculture 2022. Towards Blue Transformation. Rome, FAO. doi:10.4060/cc0461en.

Free, C. M., Cabral, R. B., Froehlich, H. E., Battista, W., Ojea, E., O'Reilly, E., ... & Gaines, S. D. (2022). Expanding ocean food production under climate change. Nature, 605(7910), 490-496. DOI: 10.1038/s41586-022-04674-5.

Foale, S., Adhuri, D., Aliño, P., Allison, E. H., Andrew, N., Cohen, P., ... & Weeratunge, N. (2013). Food security and the coral triangle initiative. *Marine Policy, 38*, 174-183. doi:10.1016/j.marpol.2012.05.033.

Fodrie, F. J., Heck Jr, K. L., Powers, S. P., Graham, W. M., & Robinson, K. L. (2010). Climate-related, decadal-scale assemblage changes of seagrass-associated fishes in the northern Gulf of Mexico. *Global Change Biology, 16*(1), 48-59. doi: 10.1111/j.1365-2486.2009.01889.x.

Font, W. F. (2003). The global spread of parasites: what do Hawaiian streams tell us?. *BioScience, 53*(11), 1061-1067. doi: 10.1641/0006-3568(2003)053[1061:TGSOPW]2.0.CO;2.

Frawley, T. H., Finkbeiner, E. M., & Crowder, L. B. (2019). Environmental and institutional degradation in the globalized economy. *Ecology and Society, 24*(1). doi:10.5751/ES-10693-240107.

Freeman, O. E. (2017). Impact of climate change on aquaculture and fisheries in Nigeria; A review. *International Journal of Multidisciplinary Research and Development, 4*(1), 53-59.

Frihy, O. E., & El-Sayed, M. K. (2013). Vulnerability risk assessment and adaptation to climate change induced sea level rise along the Mediterranean coast of Egypt. *Mitigation and Adaptation Strategies for Global Change, 18*, 1215-1237. doi:10.1007/s11027-012-9418-y.

Fujii, T. (2012). Climate change, sea-level rise and implications for coastal and estuarine shoreline management with particular reference to the ecology of intertidal benthic macrofauna in NW Europe. *Biology, 1*(3), 597-616. doi: 10.3390/biology1030597.

Gaines, S. D., Costello, C., Owashi, B., Mangin, T., Bone, J., Molinos, J. G., ... & Ovando, D. (2018). Improved fisheries management could offset many negative effects of climate change. *Science advances, 4*(8), eaao1378. doi: 10.1126/sciadv.aao1378.

Galappaththi, E. K., Ford, J. D., & Bennett, E. M. (2020). Climate change and adaptation to social-ecological change: the case of indigenous people and culture-based fisheries in Sri Lanka. *Climatic Change, 162*, 279-300. doi:10.1007/s10584-020-02716-3.

Gelwick, F. P. (1990). Longitudinal and temporal comparisons of riffle and pool fish assemblages in a northeastern Oklahoma Ozark stream. *Copeia*, 1072-1082. doi:10.2307/1446491.

Hassan, I. Musa, R.M. Latiff Azmi, M. N. Razali Abdullah, M. Yusoff, S. Z. (2023). Analysis of climate change disinformation across types, agents and media platforms. Information Development. 29: 5406–5414. doi: 10.1177/02666669221148693.

Heck N, Beck M. W., Reguero B, Pfliegner K, Ricker M, Prutz R. (2023). Global climate change risk to fisheries – A multi-risk assessment. *Marine Policy.* 148: 105404. doi:10.1016/j.marpol.2022.105404.

Hilborn, R. (2011). Future directions in ecosystem based fisheries management: a personal perspective. *Fisheries Research, 108*(2-3), 235-239. doi:10.1016/j.fishres.2010.12.030.

Hlohowskyj, I., Brody, M. S., & Lackey, R. T. (1996). Methods for assessing the vulnerability of African fisheries resources to climate change. *Climate Research, 6*(2), 97-106. doi: 10.3354/cr006097.

Hoegh-Guldberg, O. and Bruno, J. F. (2010) The impact of climate change on the world's marine ecosystems. *Science* 328, 1523–1528. doi: 10.1126/science.118993.

Holsman, K. K., Hazen, E. L., Haynie, A., Gourguet, S., Hollowed, A., Bograd, S. J., ... & Aydin, K. (2019). Towards climate resiliency in fisheries management. *ICES Journal of Marine Science, 76*(5), 1368-1378. doi:10.1093/icesjms/fsz031.

Hood, W. W., & Wilson, C. S. (2001). The literature of bibliometrics, scientometrics, and informetrics. *Scientometrics*, 52(2), 291-314. DOI: 10.1023/a:1017919924342.

Hulme, M. (2002). *Climate change scenarios for the United Kingdom*: the UKCIP02 scientific report. Tyndall Centre for Climate Mental Sciences University: School of Environmental Sciences, University of East Anglia, Norwich, UK. ISBN 0 902170 60 0.

Hutchings, J. A., & Myers, R. A. (1994). What can be learned from the collapse of a renewable resource? Atlantic cod, *Gadus morhua*, of Newfoundland and Labrador. *Canadian Journal of Fisheries and Aquatic Sciences, 51*(9), 2126-2146. doi:10.1139/f94-214.

Ibe, A. C., & Awosika, L. F. (1991). Sea level rise impact on African coastal zones. *A change in the weather: African perspectives on climate change*, 105-12.

Jaya, I., Satria, F., Nugroho, D., Sadiyah, L., Buchary, E. A., White, A. T., ... & Green, S. J. (2022). Are the working principles of fisheries management at work in Indonesia?. *Marine Policy*, *140*, 105047. doi:10.1016/j.marpol.2022.105047.

Johnson, J. E., & Welch, D. J. (2009). Marine fisheries management in a changing climate: a review of vulnerability and future options. *Reviews in Fisheries Science*, *18*(1), 106-124. doi:10.1080/10641260903434557.

Kamarudin, M. K. A., Toriman, M. E., Rosli, M. H., Juahir, H., Aziz, N. A. A., Azid, A., Zainuddin, S. F. M., & Wan, N. A. S. (2015). Analysis of meander evolution studies on effect from land use and climate change at the upstream reach of the Pahang River, Malaysia. *Mitigation and Adaptation Strategies for Global Change*. 20: 1319–1334. doi: 10.1007/s11027-014-9547-6.

Kasperski, S., DePiper, G. S., Haynie, A. C., Blake, S., Colburn, L. L., Freitag, A., ... & Weijerman, M. (2021). Assessing the state of coupled social-ecological modeling in support of ecosystem based fisheries management in the United States. *Frontiers in Marine Science*, *8*, 631400. doi:10.3389/fmars.2021.631400.

Koch, F., Wieser, W., & Niederstätter, H. (1992). Interactive effects of season and temperature on enzyme activities, tissue and whole animal respiration in roach, *Rutilus rutilus*. In *Environmental biology of European cyprinids: Papers from the workshop on 'The Environmental Biology of Cyprinids' held at the University of Salzburg, Austria, in September 1989* (pp. 73-86). Springer Netherlands. doi:10.1007/978-94-011-2544-4_8.

Kroeker, K. J., Kordas, R. L, Crim, R. N. and Singh, G. G. (2010) Meta-analysis reveals negative yet variable effects of ocean acidification on marine organisms. *Ecology Letters* 13, 1419–1434. doi: 10.1111/j.1461-0248.2010.01518.x.

Lam, V. W., Cheung, W. W., Swartz, W., & Sumaila, U. R. (2012). Climate change impacts on fisheries in West Africa: implications for economic, food and nutritional security. *African Journal of Marine Science*, *34*(1), 103-117. doi:10.2989/1814232X.2012.673294.

Lam, V. W., Cheung, W. W., Reygondeau, G., & Sumaila, U. R. (2016). Projected change in global fisheries revenues under climate change. *Scientific Reports*, *6*(1), 32607. doi:10.1038/srep32607.

Lam, V. W., Allison, E. H., Bell, J. D., Blythe, J., Cheung, W. W., Frölicher, T. L., ... & Sumaila, U. R. (2020). Climate change, tropical fisheries and prospects for sustainable development. *Nature Reviews Earth & Environment*, *1*(9), 440-454.

Last, P. R., White, W. T., Gledhill, D. C., Hobday, A. J., Brown, R., Edgar, G. J., & Pecl, G. (2011). Long-term shifts in abundance and distribution of a temperate fish fauna: a response to climate change and fishing practices. *Global Ecology and Biogeography*, *20*(1), 58-72. doi:10.1111/j.1466-8238.2010.00575.x.

Le Cornu, E., Doerr, A. N., Finkbeiner, E. M., Gourlie, D., & Crowder, L. B. (2018). Spatial management in small-scale fisheries: A potential approach for climate change adaptation in Pacific Islands. *Marine Policy*, *88*, 350-358. doi:10.1016/j.marpol. 2017.09.030.

Levine, R. M., De Robertis, A., Grunbaum, D., Wildes, S., Farley, E. V., Stabeno, P. J., & Wilson, C. D. (2023). Climate-driven shifts in pelagic fish distributions in a rapidly changing Pacific Arctic. *Deep Sea Research Part II: Topical Studies in Oceanography*, 105244. doi:10.1016/j.dsr2.2022.105244.

Leydesdorff, L., & Rafols, I. (2011). Indicators of the interdisciplinarity of journals: Diversity, centrality, and citations. *Journal of Informetrics*, *5*(1), 87-100. DOI: 10.1016/j.joi.2010.09.002.

Liu, J. M., Borazon, E. Q., & Muñoz, K. E. (2021). Critical problems associated with climate change: a systematic review and meta-analysis of Philippine fisheries research. *Environmental Science and Pollution Research*, *28*(36), 49425-49433. DOI: 10.1007/s11356-021-15712-6.

Lymer, D., Funge-Smith, S., Clausen, J., & Miao, W. (2008). Status and potential of fisheries and aquaculture in Asia and the Pacific 2008. *RAP Publication (FAO)*. ISBN 978-92-5-106109-1.

Lynch, A. J., Rahel, F. J., Limpinsel, D., Sethi, S. A., Engman, A. C., Lawrence, D. J., ... & Porath, M. T. (2022). Ecological and social strategies for managing fisheries using the Resist-Accept-Direct (RAD) framework. *Fisheries Management and Ecology*, *29*(4), 329-345. doi:10.1111/fme.12545.

Macusi, E. D., Macusi, E. S., Jimenez, L. A., and Catam-isan, J. P. (2020). Climate change vulnerability and perceived impacts on small-scale fisheries in eastern Mindanao. *Ocean Coastal Management* 189:105143. doi: 10.1016/j.ocecoaman. 2020.105143.

Macusi, E. D., Camaso, K. L., Barboza, A., & Macusi, E. S. (2021). Perceived Vulnerability and Climate Change Impacts on Small-Scale Fisheries in Davao Gulf, Philippines. *Frontiers in Marine Science*. 8: 597385. doi: 10.3389/fmars.2021. 597385.

Magness, D. R., Wagener, E., Yurcich, E., Mollnow, R., Granfors, D., & Wilkening, J. L. (2022). A multi-scale blueprint for building the decision context to implement climate change adaptation on National Wildlife Refuges in the United States. *Earth*, *3*(1), 136-156. doi:10.3390/earth3010011.

Mangano, M. C., Mieszkowska, N., Helmuth, B., Domingos, T., Sousa, T., Baiamonte, G., ... & Sarà, G. (2020). Moving toward a strategy for addressing climate displacement of marine resources: A proof-of-concept. *Frontiers in Marine Science*, *7*, 408. doi:10.3389/fmars.2020.00408.

Mamauag, S. S., Aliño, P. M., Martinez, R. J. S., Muallil, R. N., Doctor, M. V. A., Dizon, E. C., ... & Cabral, R. B. (2013). A framework for vulnerability assessment of coastal fisheries ecosystems to climate change—Tool for understanding resilience of fisheries (VA–TURF). *Fisheries Research*, *147*, 381-393. doi:10.1016/j.fishres.2013.07.007.

Mariotti, G., Fagherazzi, S., Wiberg, P. L., McGlathery, K. J., Carniello, L., & Defina, A. (2010). Influence of storm surges and sea level on shallow tidal basin erosive processes. *Journal of Geophysical Research: Oceans*, *115*(C11). doi:10.1029/2009JC005892.

Martell, S., & Stewart, I. (2014). Towards defining good practices for modeling time-varying selectivity. *Fisheries Research*, *158*, 84-95. doi:10.1016/j.fishres.2013.11.001.

Mason, R. P., Choi, A. L., Fitzgerald, W. F., Hammerschmidt, C. R., Lamborg, C. H., Soerensen, A. L., & Sunderland, E. M. (2012). Mercury biogeochemical cycling in the ocean and policy implications. *Environmental research*, *119*, 101-117. doi:10.1016/j.envres.2012.03.013.

McCarthy, J. J., Canziani, O. F., Leary, N. A., Dokken, D. J., & White, K. S. (Eds.). (2001). *Climate change 2001: impacts, adaptation, and vulnerability: contribution of Working Group II to the third assessment report of the Intergovernmental Panel on Climate Change* (Vol. 2). Cambridge University Press. ISBN 0 521 80768 9.

McClanahan, T. R. (2010) Effects of fisheries closures and gear restrictions on fishing income in a Kenyan coral reef. *Conservation Biology* 24, 1519–1528. doi:10.1111/j.1523-1739.2010.01530.x.

Melzner, F., Gutowska, M. A., Langenbuch, M., Dupont, S., Lucassen, M., Thorndyke, M. C., ... & Pörtner, H. O. (2009). Physiological basis for high CO 2 tolerance in marine ectothermic animals: pre-adaptation through lifestyle and ontogeny?. *Biogeosciences*, *6*(10), 2313-2331. doi:10.5194/bg-6-2313-2009.

Mohanty, B., Vivekanandan, E., Mohanty, S., Mahanty, A., Trivedi, R., Tripathy, M., & Sahu, J. (2017). The impact of climate change on marine and inland fisheries and aquaculture in India. *Climate change impacts on fisheries and aquaculture: a global analysis*, *2*, 569-601. doi: 10.1002/9781119154051.ch17.

Mondal, S. K., Huang, J., Wang, Y., Su, B., Kundzewicz, Z. W., Jiang, S., ... & Jiang, T. (2022). Changes in extreme precipitation across South Asia for each 0.5°C of warming from 1.5°C to 3.0°C above pre-industrial levels. *Atmospheric Research*, *266*, 105961. doi:10.1016/j.atmosres.2021.105961.

Monnet, G., Corse, E., Archambaud-Suard, G., Grenier, R., Chappaz, R., & Dubut, V. (2022). Growth variation in the endangered fish *Zingel asper*: Contribution of substrate quality, hydraulics, prey abundance, and water temperature. *Aquatic Conservation: Marine and Freshwater Ecosystems*, *32*(7), 1156-1170. doi:10.1002/aqc.3818.

Newton, K., Cote, I. M., Pilling, G. M., Jennings, S. and Dulvy, N. K. (2007) Current and future sustainability of island coral reef fisheries. *Current Biology* 17, 655–658. doi: 10.1016/j.cub.2007.02.054.

Olawumi, T. O., & Chan, D. W. (2018). A scientometric review of global research on sustainability and sustainable development. *Journal of Cleaner Production*, *183*, 231-250. doi: 10.1016/j.jclepro.2018.02.162.

Pethick, J.S.; Crooks, S. Development of a coastal vulnerability index: A geomorphological perspective. *Environment Conservation.* 2000, 27, 359–367. doi: 10.1017/S0376892900000412.

Plagányi, É. E. (2007). *Models for an ecosystem approach to fisheries*. FAO Fisheries Technical Paper, 477. ISBN 978-92-5-105734-6.

Pikitch, E. K., Santora, C., Babcock, E. A., Bakun, A., Bonfil, R., Conover, D. O., ... & Sainsbury, K. J. (2004). Ecosystem-based fishery management. *Science*, *305*(5682), 346-347. DOI: 10.1126/science.1098222.

Poloczanska, E. S., Burrows, M. T., Brown, C. J., García Molinos, J., Halpern, B. S., Hoegh-Guldberg, O., ... & Sydeman, W. J. (2016). Responses of marine organisms to climate change across oceans. *Frontiers in Marine Science*, 62. DOI: 10.3389/fmars.2016.00062.

Pons, M., Branch, T. A., Melnychuk, M. C., Jensen, O. P., Brodziak, J., Fromentin, J. M., ... & Hilborn, R. (2017). Effects of biological, economic and management factors on tuna and billfish stock status. *Fish and Fisheries*, *18*(1), 1-21. doi:10.1111/faf.1216.

Porobic, J., Fulton, E. A., Frusher, S., Parada, C., Haward, M., Ernst, B., & Stram, D. (2018). Implementing ecosystem-based fisheries management: lessons from Chile's experience. *Marine Policy*, *97*, 82-90. doi:10.1016/j.marpol.2018.08.037.

Pörtner, H. O., & Knust, R. (2007). Climate change affects marine fishes through the oxygen limitation of thermal tolerance. *Science*, *315*(5808), 95-97. doi: 10.1126/science.1135471.

Post, J. R., Sullivan, M., Cox, S., Lester, N. P., Walters, C. J., Parkinson, E. A., ... & Shuter, B. J. (2002). Canada's recreational fisheries: the invisible collapse?. *Fisheries*, *27*(1), 6-17. doi:10.1577/1548-8446(2002)027<0006:CRF>2.0.CO;2.

Pozo Buil, M., Jacox, M. G., Fiechter, J., Alexander, M. A., Bograd, S. J., Curchitser, E. N., ... & Stock, C. A. (2021). A dynamically downscaled ensemble of future projections for the California current system. *Frontiers in Marine Science*, *8*, 612874. doi:10.3389/fmars.2021.612874.

Quynh, C. N. T., Schilizzi, S., Hailu, A., & Iftekhar, S. (2017). Territorial Use Rights for Fisheries (TURFs): State of the art and the road ahead. *Marine Policy*, *75*, 41-52. doi:10.1016/j.marpol.2016.10.004.

Radhakrishnan, S., Erbis, S., Isaacs, J. A., & Kamarthi, S. (2017). Novel keyword co-occurrence network-based methods to foster systematic reviews of scientific literature. *PloS one*, *12*(3), e0172778. doi:10.1371/journal.pone.0172778.

Rice, J. C., & Garcia, S. M. (2011). Fisheries, food security, climate change, and biodiversity: characteristics of the sector and perspectives on emerging issues. *ICES Journal of Marine Science*, *68*(6), 1343-1353. doi:10.1093/icesjms/fsr041.

Rijnsdorp, A. D., Peck, M. A., Engelhard, G. H., Möllmann, C., & Pinnegar, J. K. (2009). Resolving the effect of climate change on fish populations. *ICES journal of marine science*, *66*(7), 1570-1583. doi:10.1093/icesjms/fsp056.

Ruckelshaus, M., Doney, S. C., Galindo, H. M., Barry, J. P., Chan, F., Duffy, J. E., ... & Talley, L. D. (2013). Securing ocean benefits for society in the face of climate change. *Marine Policy*, *40*, 154-159. doi:10.1016/j.marpol.2013.01.009.

Rykaczewski, R. R., & Checkley Jr, D. M. (2008). Influence of ocean winds on the pelagic ecosystem in upwelling regions. *Proceedings of the National Academy of Sciences*, *105*(6), 1965-1970. doi:10.1073/pnas.0711777105.

Salgueiro-Otero, D., Barnes, M. L., & Ojea, E. (2022). Climate adaptation pathways and the role of social-ecological networks in small-scale fisheries. *Scientific Reports*, *12*(1), 1-13. doi:10.1038/s41598-022-18668-w.

Sax, D. F., Stachowicz, J. J., Brown, J. H., Bruno, J. F., Dawson, M. N., Gaines, S. D., ... & Rice, W. R. (2007). Ecological and evolutionary insights from species invasions. *Trends in ecology & evolution*, *22*(9), 465-471. doi:10.1016/j.tree.2007.06.009.

Schirripa, M. J., Goodyear, C. P., & Methot, R. M. (2009). Testing different methods of incorporating climate data into the assessment of US West Coast sablefish. *ICES Journal of Marine Science*, *66*(7), 1605-1613. doi:10.1093/icesjms/fsp043.

Schuurman, G. W., Cole, D. N., Cravens, A. E., Covington, S., Crausbay, S. D., Hoffman, C. H., ... & O'Malley, R. (2022). Navigating ecological transformation: Resist–accept–direct as a path to a new resource management paradigm. *BioScience*, *72*(1), 16-29. doi:10.1093/biosci/biab067.

Shaffril H. A. M., Samah A. A., D'Silva JL. (2017). Adapting towards climate change impacts: Strategies for small-scale fishermen in Malaysia. *Marine Policy*. 81: 196-201. doi:10.1016/j.marpol.2017.03.032.

Suh, D., & Pomeroy, R. (2020). Projected economic impact of climate change on marine capture fisheries in the Philippines. *Frontiers in Marine Science*, *7*, 232. doi:10.3389/fmars.2020.00232.

Szuwalski, C. S., & Hollowed, A. B. (2016). Climate change and non-stationary population processes in fisheries management. *ICES Journal of Marine Science*, *73*(5), 1297-1305. doi:10.1093/icesjms/fsv229.

Techera, E. J. (2018). Supporting blue economy agenda: fisheries, food security and climate change in the Indian Ocean. *Journal of the Indian Ocean Region*, *14*(1), 7-27. doi:10.1080/19480881.2017.1420579.

Thirukanthan, C. S., Azra, M. N., Lananan, F., Sara', G., Grinfelde, I., Rudovica, V., ... & Burlakovs, J. (2023). The Evolution of Coral Reef under Changing Climate: A Scientometric Review. *Animals*, *13*(5), 949.

Thorpe, A., Reid, C., van Anrooy, R., Brugere, C., & Becker, D. (2006). Asian development and poverty reduction strategies: integrating fisheries into the development discourse. *Food Policy*, *31*(5), 385-400. doi:10.1016/j.foodpol.2005.09.007.

Tian, S., Chen, Y., Chen, X., Xu, L., & Dai, X. (2009). Impacts of spatial scales of fisheries and environmental data on catch per unit effort standardisation. *Marine and freshwater research*, *60*(12), 1273-1284. doi: 10.1071/MF09087.

Tommasi, D., Stock, C. A., Pegion, K., Vecchi, G. A., Methot, R. D., Alexander, M. A., & Checkley Jr, D. M. (2017). Improved management of small pelagic fisheries through seasonal climate prediction. *Ecological Applications*, *27*(2), 378-388. doi: 10.1002/eap.1458.

Townsend, H., Harvey, C. J., DeReynier, Y., Davis, D., Zador, S. G., Gaichas, S., ... & Kaplan, I. C. (2019). Progress on implementing ecosystem-based fisheries management in the United States through the use of ecosystem models and analysis. *Frontiers in Marine Science*, *6*, 641. doi:10.3389/fmars.2019.00641.

Waltman, L. (2016). A review of the literature on citation impact indicators. *Journal of Informetrics*, *10*(2), 365-391. DOI: 10.1016/j.joi.2016.02.007.

Wedemeyer, G. (1996). *Physiology of fish in intensive culture systems*. Springer Science & Business Media.

Worm, B., Hilborn, R., Baum, J. K., Branch, T. A., Collie, J. S., Costello, C., ... & Zeller, D. (2009). Rebuilding global fisheries. *Science*, *325*(5940), 578-585. doi: 10.1126/science.1173146.

Velumani, T., P. S. Ananthan, P. Krishnan and Lloyd Chrispin, C. (2019). Research on Climate Change and Fisheries: A Scientometric Analysis. *International Journal of Current Microbiology and Applied Sciences*. 8: 336-345. doi:10.20546/ijcmas. 2019.808.038.

Vousdoukas, M. I., Clarke, J., Ranasinghe, R., Reimann, L., Khalaf, N., Duong, T. M., ... & Simpson, N. P. (2022). African heritage sites threatened as sea-level rise accelerates. *Nature Climate Change*, *12*(3), 256-262. doi:10.1038/s41558-022-01280-1.

Xiao, Q., & Huang, S. (2018). Bibliometric analysis of impacts of climate change on fisheries. *Journal of Shanghai Ocean University*, *27*(2), 304-310. ISSN: 1674-5566.

Yang, H., Shao, X., & Wu, M. (2019). A review on ecosystem health research: A visualization based on CiteSpace. *Sustainability*, *11*(18), 4908. doi: 10.3390/su11184908.

Zhong, Z., Zheng, L., Luo, Z., Li, S., & Yang, Y. (2019). Invariance matters: Exemplar memory for domain adaptive person re-identification. In *Proceedings of the IEEE/CVF Conference on Computer Vision and Pattern Recognition* (pp. 598-607).

Zou, M., Tian, W., Zhao, J., Chu, M., & Song, T. (2022). Quinolone antibiotics in sewage treatment plants with activated sludge treatment processes: a review on source, concentration and removal. *Process Safety and Environmental Protection*, *160*, 116-129. doi: 10.1016/j.psep.2022.02.013.

Chapter 4

A Scientometric Analysis on the Impacts of Climate Change on Molluscs

Roslizawati binti Ab. Lah[1,*]
Selma Bencedira[2]
Murni Nur Islamiah Kassim[3]
Mohamad Nor Azra[3,4]
Thirukanthan Chandra Segaran[1]
Mohd Iqbal Mohd Noor[5,6]
Hidir Ariffin[7]
Zulhisyam Abdul Kari[8]
Walter G. Bottje
and Wan Mohd Rauhan Wan Hussin[1,†]

[1]Universiti Malaysia Terengganu (UMT), Kuala Nerus, Terengganu, Malaysia
[2]Laboratory of LGE, Department of Process Engineering, University Badji-Mokhtar Annaba, Annaba, Algeria
[3]Institute of Climate Adaptation and Marine Biotechnology (IMB), Universiti Malaysia Terengganu (UMT), Terengganu, Malaysia
[4]Research Center for Marine and Land Bioindustry, Earth Sciences and Maritime Organization, National Research and Innovation Agency (BRIN), Pemenang, West Nusa Tenggara, Indonesia
[5]Institute for Biodiversity and Sustainable Development, Universiti Teknologi MARA (UiTM), Shah Alam, Selangor, Malaysia
[6]Faculty of Business Management, Universiti Teknologi MARA (UiTM) (Pahang), Raub, Pahang, Malaysia

[*] Corresponding Author's Email: ros_ablah@umt.edu.my.
[†] Corresponding Author's Email: rauhan@umt.edu.my.

In: Marine Life in Changing Climates
Editors: M. Nor Azra, T. Chandra Segaran, G. Téllez-Isaías et al.
ISBN: 979-8-89113-404-1

[7]Higher Institution Centre of Excellence (HICoE), Institute of Tropical Aquaculture and Fisheries (AKUATROP), Universiti Malaysia Terengganu, Terengganu, Malaysia
[8]Department of Agricultural Science, Faculty of Agro-Based Industry, Universiti Malaysia Kelantan, Jeli, Kelantan, Malaysia
[9]Department of Poultry Science, University of Arkansas, Fayetteville, AR, USA

Abstract

Shellfish serve as vital protein sources for humans, representing some of the primary aquacultured food resources globally. Species like molluscs are predominantly cultured in coastal regions, though there's a growing trend towards offshore mollusc farming. Both coastal and deep-sea habitats are increasingly affected by the repercussions of climate change. Elevated temperatures warm the ocean, subsequently impacting ecological dynamics and the organisms within these ecosystems. Therefore, the reliance of shellfish culture on aquatic ecosystems may affect this protein source production. This study reviews an emerging research frontier by identifying significant impacts of climate change on the shellfish, particularly molluscs. The Web of Science Core Collection (WOSCC) database was used as a main proxy to extract the bibliometric information and CiteSpace software was used to analyze the scientometric dataset. Data was generated from WOSCC from 1970 until December 31, 2021. A total of 28,061 articles were generated, with inferential statistics from a descriptive dataset showing collaborative networks between authors, institutions and countries. Meanwhile the citation data also indicated that keywords such as oxidative stress, growth, temperature, rat, hypoxia, and calcium are among the most used keywords. When examining the prevailing trends in leading publications on this topic, ocean acidification emerges as a primary concern, as corroborated by recent research. Recognizing the significance of this issue and the observed research gap concerning the effects of climate change on molluscs, this study endeavors to bridge this void. To the best of the author's understanding, this is the inaugural paper employing scientometric analysis in the context of mollusc-related research on climate change. Data retrieved from the WOSCC database indicates a consistent trajectory of publications, highlighting the burgeoning potential of this area as an emergent and prospective field of study.

Keywords: climate change, food resources, shellfish, molluscs, scientometric

1. Introduction

Elements of climate change, including global warming, the greenhouse effect, carbon dioxide (CO2) concentrations, and oceanic temperatures, are intricately interconnected, influencing shifts in marine ecosystems (Bierly & Kingsford, 2009; Thirukantan et al., 2023). For instance, atmospheric CO2 levels have escalated from 380 ppm in 2004 to 420 ppm in 2023 (NOAA, 2021; Hale et al., 2011). Elevations of CO_2 are attributed to human activities including the combustion of fossil fuels, deforestation and agriculture are just a few examples that have contributed to an increase in atmospheric CO_2 levels which lead to global warming (CSIRO, 2006; Nunes, 2023). Absorption of CO_2 into the ocean resulting in the decrease of pH level of 0.3–0.4 units over this century (IPCC, 2007). Additionally, it has been observed that between 1880 and 2012, the temperature of the ocean showed an increase of about 0.85 [0.65 to 1.06] °C.'

Climate change poses a significant challenge to the sustainability and progression of fisheries in tropical regions, particularly for molluscs. Historical data on sea surface temperatures in these tropical zones, which boast some of the most extensive coastal areas, suggests potential impacts on mollusc species. The potential ramifications for marine organisms that calcify, especially molluscs, are profound. This is supported by several recent studies and meta-analyses examining the influence of global climate change on marine life (Guinotte et al., 2008; Henriks et al., 2009; Byrne, 2011; Ablah et al., 2018). Since mollusk shells are often constructed from various calcium carbonate compounds, they serve as one of the greatest models for evaluating climate change because they are extremely sensitive to environmental changes (Kroeker et al., 2013; Fortunato & Hric, 2016; Sowa et al., 2019). Previous research indicates that molluscs, especially those with heavily calcified shells, are more vulnerable to ocean acidification compared to crustaceans (Wittmann, 2013). In addition, molluscs have been effectively used as a trustworthy paleoclimate proxy that might provide historical data on oceanic variability to comprehend any future changes that may take place (Prentice, 2011; Schöne, 2013). Therefore, using molluscs in climate change studies will yield important information on changes in our oceans.

Indeed, molluscs are not exclusive to marine habitats. They can be cultivated across tropical, marine, and temperate environments using a variety of farming methods. This includes extensive recirculation systems for abalones and semi-intensive line and raft cultures for mussels and scallops. They also contribute in a variety of other ways to the economy and society, such as via food consumption, jewelry, and decoration (Wittmann, 2013). Currently, there are several studies conducted on the impact of climate change on molluscs, mostly in review-based studies (Kroeker et al., 2010; Tan & Zheng, 2020). However, studies related to the impact of climate change on molluscs, specifically in terms of bibliometric-based analysis is still lacking.

Bibliometric analysis is popularly known as "research of research" and among one of the studies that use statistical methods to identify any relationship between published literature and meta information (Hou et al., 2019; Mejia et al., 2021). The application of bibliometrics provides interesting insights into global selected research, such as climate change associated with major aquatic molluscs in the world. Meanwhile, CiteSpace is among one of the most important visualization analysis software that is used for the bibliometric related studies (Miao et al., 2021; Chen, 2020).

This chapter adopted the bibliometric techniques through the scientometric analysis to conduct literature evaluations of molluscs impacted by climate change. The study is expected to provide insight on the past, present and future situation of climate drivers on molluscs, and how this will influence the future directions on funding and research agenda. Information from the present study may be beneficial to various stakeholders including climate change researchers, academic and research institutions, policy makers and industries related to mollusc farming to better understand the past and present impacts of climate change towards this calcifying organism. This then may provide support to the aquaculture industry in preparing any emergency plan in the future.

In summary, up to now, a huge number of studies have focused on the impact of climate change on molluscs. However, to the best of our knowledge, the current study is the first public research on the bibliometric examination of the effects of climate change on significant molluscs worldwide. The following inquiries particularly interest us: (i) What are the current publication output trends? (i.e., publication number, number of authors, affiliation involved, published journal)? (ii) Which countries focus on climate change and molluscs research and development? and is there any connection between them?, (iii) Which authors are most prominent in these fields?, (iv) Which topics and clusters are most prevalent, and how has their network changed over time?, (v) What influential publications and keywords are relevant to these fields?

2. Materials and Methods

2.1. The Search Strategy

Generally, numerous search engines, such as Scopus, Web of Science (WOS), or PubMed, can produce bibliometric data (Minasny et al., 2013). However, only a few databases contain citation or co-citation dataset (Chen, 2020). In the present study, the WOS Core Collection (WOSCC) database was used to perform the available literature from its inception until December 31, 2021. WOS was chosen for two main reasons: the first one is concerning CiteSpace, the software is limited to only one database that can be incorporated in one-time process. The second one is the fact that it offered the citation data in more details compared to other databases such as Scopus or PubMed.

The keywords for articles searching incorporated two fields that defined the search step and they were connected by a conjunction of Boolean operator, the AND option. The first search field was predetermined for keywords related to Mollusca. The keywords that are related to the molluscs were: molluscs, abalone, winkles, conchs, oysters, mussels, scallops, pectens, clams, strombus, cockles, arkshells, gastropod and whelks. The various terms of Mollusca used in this study were chosen because they are the most common terms used by FAO 2022 and can be found throughout the world.

The second search field was predetermined for keywords related to climate change and they were based on the recently published article by (Azra et al., 2022). Briefly, the keywords with the asterisk symbol (*) are: (climat*) OR ("global warm*") OR ("seasonal* variat*") OR ("extrem* event*") OR ("environment* variab*") OR ("anthropogenic effect*") OR ("stres*") OR ("greenhouse effect*") OR ("sea level ris*") OR (erosio*) OR ("agricult* runoff") OR ("weather* variab*") OR ("weather* extrem*") OR ("extreme* climat*") OR ("environment* impact*") OR ("environment* chang*") OR ("anthropogenic stres*") OR ("temperature ris*") OR ("temperature effect*") OR ("warm* ocean") OR ("sea surface* temperat*") OR (heatwav*) OR (acidific*) OR (hurrican*) OR (El Nino) OR ("El Nino") OR ("La Nina") OR (La Nina) OR (drought*) OR (flood*) OR ("high precipit*") OR ("heavy rainfall*") OR ("CO2 concentrat*") OR ("melt* of the glacier*") OR ("melt* ice*") OR ("therm* stress*") OR (drought) OR (hypoxia)).

The synonyms for both molluscs and global warming were applied in the search topic that includes all the article title, abstract, author keywords, and Keywords Plus. After applying the search fields, a preliminary analysis was carried out to exclude articles that do not correspond to the research aim. This includes articles under several WOSCC categories including Neurosciences, Immunology and Parasitology. Articles in any other language other than English were also excluded. After filtering the list, bibliometric data of the retrieved publications was exported in the form of Plain Text File (.txt) from WOSCC.

2.2. Data Analysis

The CiteSpace software, available freely, was employed to visualize the primary data pertinent to this subject. This data was cataloged as "Full Record and Cited References" in WOSCC. To discern the trends, evolution, frontier research, and focal points of publications regarding the impact of climate change on molluscs, CiteSpace was selected for a gamut of citation and collaboration analyses. These analyses encompassed Author citation analysis, country networks, document citation cluster analysis, and networks, as well as identifying the Strongest Citation Bursts, often referred to as burstiness in documents and keywords.

CiteSpace incorporates several data-cleaning tools that enhance the bibliographic data used in analyses. For instance, it has a "deduplication" feature that identifies and omits redundant entries from a dataset, thereby curtailing the likelihood of analytical bias. Additionally, CiteSpace facilitates the merging of multiple datasets into one consolidated file and ensures data consistency across various sources through its standardization tool. Collectively, these data refinement tools augment the integrity and precision of the data processed in CiteSpace, which consequently yields more dependable and insightful outcomes.

2.3. Concepts and Metrics

Some important terms are used in scientometric studies. These terms should be explained for a better understanding, here are the most used:

2.3.1. Betweenness Centrality

Within a network, every node possesses a specific betweenness centrality. This metric evaluates the likelihood that a randomly selected shortest path within the network will traverse that particular node. The term "betweenness" refers to the likelihood that a node with a high betweenness centrality will be located between two sizable communities or sub-networks. A node with a strong betweenness centrality is indicated in CiteSpace by a purple ring.

2.3.2. Citation Burst

A burst is an increase in frequency of a specific sort of occurrence, such as an increase in the number of citations for a publication that won the Nobel Prize. CiteSpace, for instance, provides burst detection for the following events: Single or multiple-word phrases taken from a publication's title, abstract, or other passages, as well as historical citation counts for cited sources. The quantity of publications by an author, an organization, or a nation, as well as the frequency of keyword appearances through time.

2.3.3. Co-citations and Co-Occurrences

This pertains to the concurrent citation of two references within a third article. Conventionally, two references are deemed co-cited if they are jointly referenced within this third article. While the entirety of the third article's content can serve as a context, the scope can be narrowed down to specific sections, paragraphs, or terms. Citing a reference can be done for a variety of reasons and for a wide range of objectives. The method a reference has been referenced, nevertheless, might also be used to reference an underlying idea.

2.3.4. Sigma

In a network of cited references, a node's significance is quantified by a metric termed "sigma." This metric highlights nodes that are structurally pivotal, particularly those demonstrating citation burstness, which indicates a rapid surge in citations. Sigma is effective for identifying works that are obviously attracting attention and could be significant.

3. Results

Recently, molluscs have gained prominence as model organisms in climate change research. Investigations related to the impact of climate change on molluscs were restricted to original research articles sourced from the Web of Science Core Collection databases, spanning from 1970 up to December 31, 2021 (as illustrated in Figure 1). After a stagnant number of publications in the first 20 years, the trend started a gradual increase from 1990 to 2010 with an increase of approximately 500 records over the course of 10 years. The rapid increase was evidenced after that with a two-fold increase i.e., approximately 1,000 new articles were recorded within 10 years.

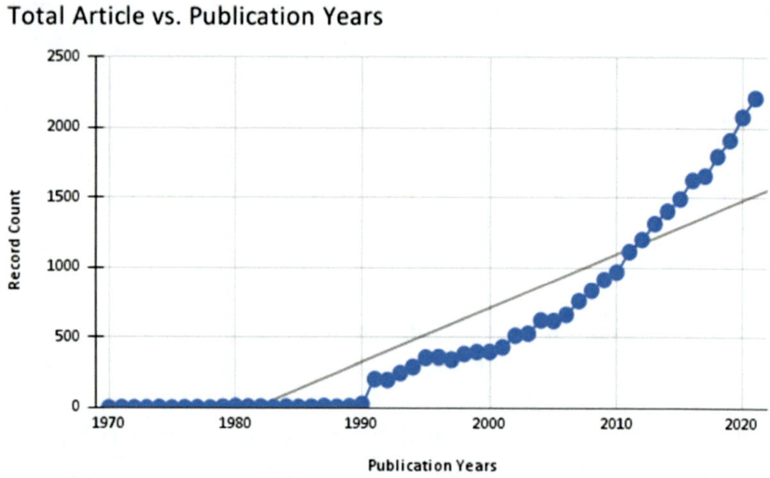

Figure 1. Total number of publications over years for the domain of climate change associated with the molluscs group.

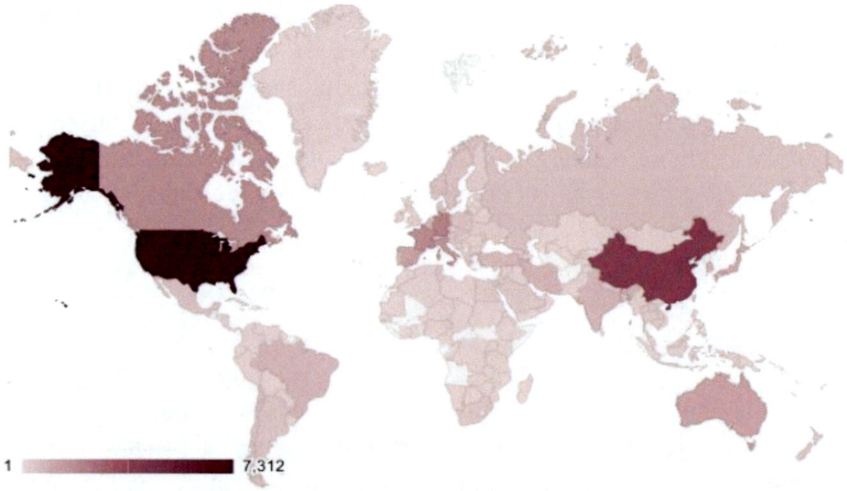

Figure 2. The geographical distribution of research on climate change and molluscs is represented by varying shades of magenta. Darker shades of magenta highlight countries with the highest number of publications, while lighter shades indicate countries with fewer publications in this domain.

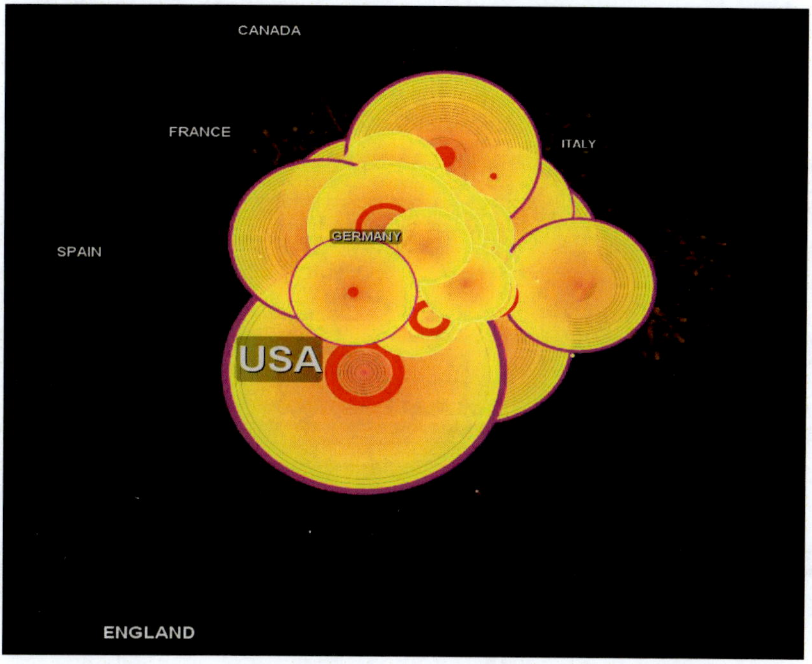

Figure 3. The network of countries, as visualized through the CiteSpace software analysis, depicts the most influential nations in climate change and molluscs research. Larger circles, marked by a pink ring around the node, represent countries with the greatest influence in this domain.

A total of 169 countries have published at least one article in the research domain of climate change and molluscs. Importantly, the mentioned country corresponds to the affiliation of the author(s). The USA markedly leads in this arena, registering the highest number of publications (as depicted in Figure 2). The top three countries in terms of publication volume are the USA, China, and Germany, collectively accounting for 47% of publications. Other significant contributors include Canada, Australia, Brazil, and a substantial portion of European nations.

Based on citation analysis by country, the USA, England, and Canada emerge as the most influential nations concerning the research theme of climate change and molluscs. Germany, Italy, France, and Spain have also made notable contributions, as illustrated in Figure 3.

England's prominence in the citation metrics can be largely attributed to L. Bayne from Plymouth Marine Laboratory, England, who emerged as the most cited author in the field, as per the Web of Science Core Collection database. Other authors with significant contributions in this domain include EG. Bligh, RJA. Jathompson, AD. Ansell, JH. Connell, RR. Sokal, CJ. Bayne, and RC. Newell, as showcased in Figure 4.

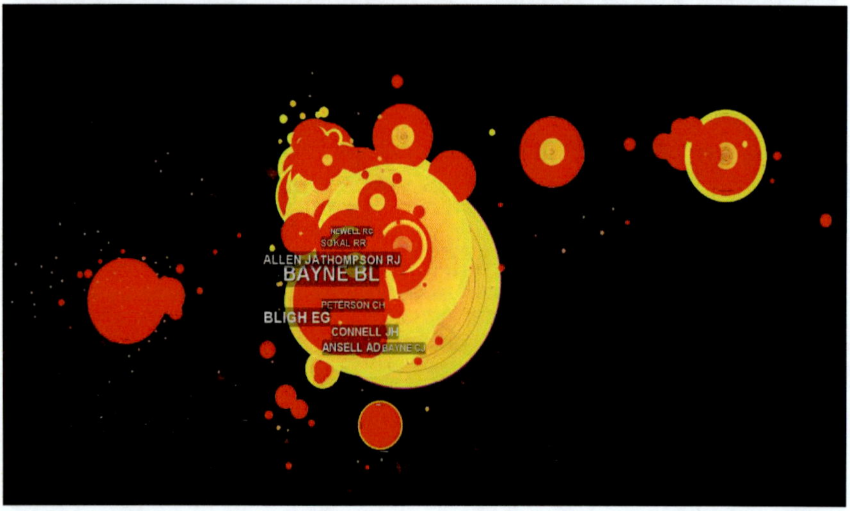

Figure 4. Author citation analysis, with bigger font, indicates the most influential authors in the current field of research, based on the WOS Core Collection database.

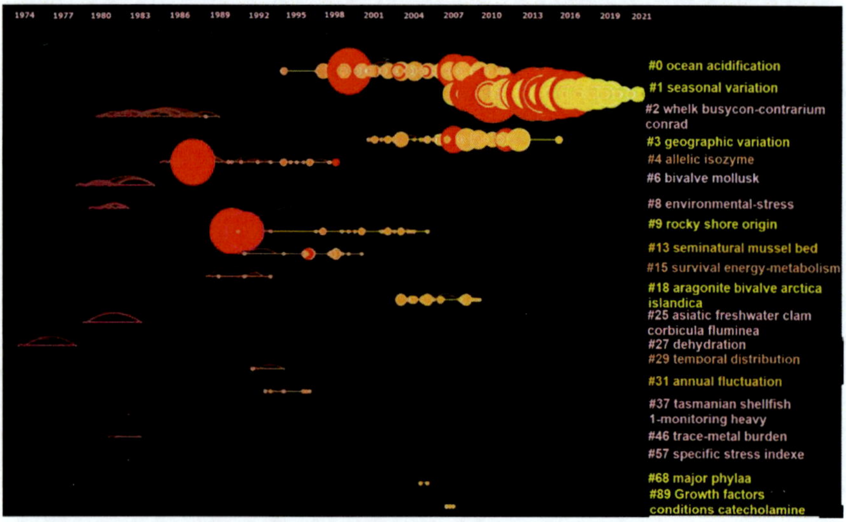

Figure 5. The summary of identified documents with cluster labels were generated from CiteSpace and were divided into 20 co-citation clusters.

Figure 6. The networks evolution of between published articles in the climate change and molluscs group published documents.

CiteSpace software generated 20 co-citation clusters under climate change and mollusk domain (Figure 5). This timeline cluster shows the attention the study received by the researchers in relation to the keywords extracted from the articles. This domain recorded a singularity of dominance in recent years where 'seasonal variation' was the only cluster with a substantial burst in citation starting from 2013 to 2021. Another two clusters namely 'ocean acidification' and 'geographic variation' recorded a fair share of burstiness despite being prematurely ended circa 2012 and 2013 respectively.

The progression of document networks is evident in the published articles relating to climate change and the molluscs group, as visualized in Figure 6. The analysis indicates that the subsequent paper by Hawkins et al., (Hawkins et al., 1986) stands as the most pivotal document in this domain. Furthermore, we also scrutinized the top 10 seminal publications, as sourced from the Web of Science Core Collection database, which are detailed in Table 1 and elaborated upon in the Discussion section.

Table 1. The impactful publications (top 10) based Strongest Citation Bursts, referred to the Web of Science Core Collection database

References	Year	Strength	Begin	End	1970 – 2022
Zhang et al.,	2012	80.57	2013	2017	
Gazeau et al.,	2013	61.18	2014	2018	
Kroeker et al.,	2013	60.72	2014	2018	
Regoli and Giuliani	2014	47.34	2015	2019	
Waldbusse et al.,	2015	40.11	2016	2020	
Thomsen et al.,	2013	39.42	2014	2018	
Ries et al.,	2009	38.02	2010	2014	
Doney et al.,	2009	36.68	2009	2014	
Fabry et al.,	2008	33.48	2009	2013	
Sokal and Rohlf	1995	31.7	1995	1998	

Table 2. Top 10 most influential keywords based Strongest Citation Bursts, referred to the Web of Science Core Collection database

Keywords	Year	Strength	Begin	End	1970 - 2021
rat	1970	66.4	1991	2009	
hypoxia	1970	61.22	1991	2007	
Calcium	1970	60.66	1991	2006	
potassium channel	1970	55.3	1991	2007	
Oxygen	1970	52.58	1993	2008	
patch clamp	1970	51.1	1991	2001	
intracellular ph	1970	46.01	1992	2002	
ion channel	1970	44.74	1991	2006	
Cadmium	1970	42.87	1991	2006	
Mytilus edulis l	1970	41.82	1988	2002	

Figure 7. The distribution of the keywords on the domain of climate change and mollusc group.

The trajectory of research on this topic has witnessed a shift between 1995 and 2015 (Table 1). During the 1990s, historical analyses of marine resource alterations demonstrated that both the fisheries and aquaculture sectors were dynamic, adeptly managing change. As the 2000s dawned, research gravitated towards the examination of water acidification and its ensuing environmental consequences. This trend, centered on acidification studies, persisted well into the early 2010s. Post-2010, the research emphasis largely transitioned to themes such as oxidative stress, genetic traits, and the ramifications of climate change on aquatic communities, with a keen interest in understanding their adaptive responses.

Table 2 presents an analysis of the most influential keywords based on Strongest Citation Bursts, as sourced from the Web of Science Core Collection database.

With the exception of the term "Mytilus edulis l," which first made its appearance in our research in 1988, all other terms, namely: "rat," "hypoxia," "calcium," "potassium channel," "patch clamp," "ion channel," and "cadmium," were introduced in 1991. Only the keywords "intracellular pH" and "oxygen" began being utilized from 1992 onward. Notably, the use of the terms "patch clamp," "intracellular pH," and "Mytilus edulis l" ceased post-

2002. Meanwhile, the remaining keywords saw a decline in usage between the years 2006 to 2009, as detailed in Table 2.

Meanwhile, Figure 7 showed the most keywords used in the "Title, Abstract & Keywords" in the WOSCC "Topic" selection, in which keywords "rat" "hypoxia," "calcium" are among the top 3 keywords in the analysis while "potassium channel," "oxygen," "patch clamp," "intracellular ph," "ion channel," "cadmium" and "*Mytilus edulis* l" are among the top 10 keywords of our analysis.

4. Discussion

4.1. Relevance of the Results and Trend Evolution Over Time

All natural water, including seawater, contains a variety of dissolved chemical compounds, such as calcium carbonate ($CaCO_3$) (Moras et al., 2022). It depicts the primary component of molluscan shells and was made when water was dissolved with atmospheric CO_2. In fact, the reaction mechanism (Equations 1-4) through which CO_2 dissolves in seawater is the calco-carbonic balance. Equation 5 states that the final limestone is created when the carbonate ions produced combine with the calcium ions. (Mitchell et al., 2010).

$$CO_2 \rightarrow H_2CO_3 \quad (1)$$

$$H_2CO_3 \rightarrow HCO_{3-} + H^+ \quad (2)$$

$$HCO2^- \rightarrow H^+ + CO_3 1^{2-} \quad (3)$$

$$H_2O \rightarrow HO^- H^+ \quad (4)$$

$$CO_3 1^{2-} + Ca_2 1^+ \rightarrow CaCO_3 \quad (5)$$

Marine biodiversity is under mounting threat owing to the unprecedented pace of global warming observed in recent decades, coupled with projections for the imminent future (CSIRO, 2006). As highlighted by the US Environmental Protection Agency (USEPA), there has been a discernible uptick in sea surface temperatures throughout the 20th century, a trend that persists. The USEPA further notes that the Earth's mean surface temperature has increased approximately 1°C since the late 19th century, with a significant portion of this warming transpiring in recent decades (US EPA, 2022). The sea surface temperature has increased steadily over the past three decades more than at any other time since reliable measurements first began in 1880. While there was a noticeable decline in sea surface temperature from 1880 to 1910, water temperature essentially captures the thermal state of water at a specific point in time and location. This metric has been independently computed by four distinct entities employing various methodologies. These include:

1. The National Oceanic and Atmospheric Administration of the United States (NOAA) (Boyer, 2022)
2. The Commonwealth Scientific and Industrial Research Organization of Australia (CSIRO) (Hendriks, 2010)
3. The Institute of Atmospheric Physics of China (IAP) (Ishii et al., 2017)
4. The Meteorological Research Institute of the Japan Meteorological Agency (PRI/MA) (Cheng et al., 2022).

Even if an alternative reference period were chosen, the temporal trajectory of the data would remain consistent. Based on the findings from these four studies, there has been a consistent warming trend in the upper 700 meters of the oceans since 1955, as illustrated in Figure 8. Moreover, Figure 8b underscores that, upon examining the upper 2,000 meters, all analyses reveal an even more pronounced warming of ocean waters. Such trends highlight the phenomenon wherein the heat absorbed by the surface waters gradually permeates and affects even the deeper oceanic strata.

Changes in species distributions are one of the warming's most obvious effects. Marine organisms that cannot adapt to the changing temperature of the water because of global warming migrate to cooler water areas or perish (Tan & Zheng, 2020).

Consequently, many researchers studied the relationship between climate change and marine organisms, specifically molluscs [Mitchell et al., 2010]. Molluscs are essential ecosystem engineers because they help to structure aquatic bottom habitats and offer a range of other species habitat, protection, and food (Moras et al., 2022). The importance of mollusks to humans historically and their current economic significance across the globe (Azra et al., 2022).

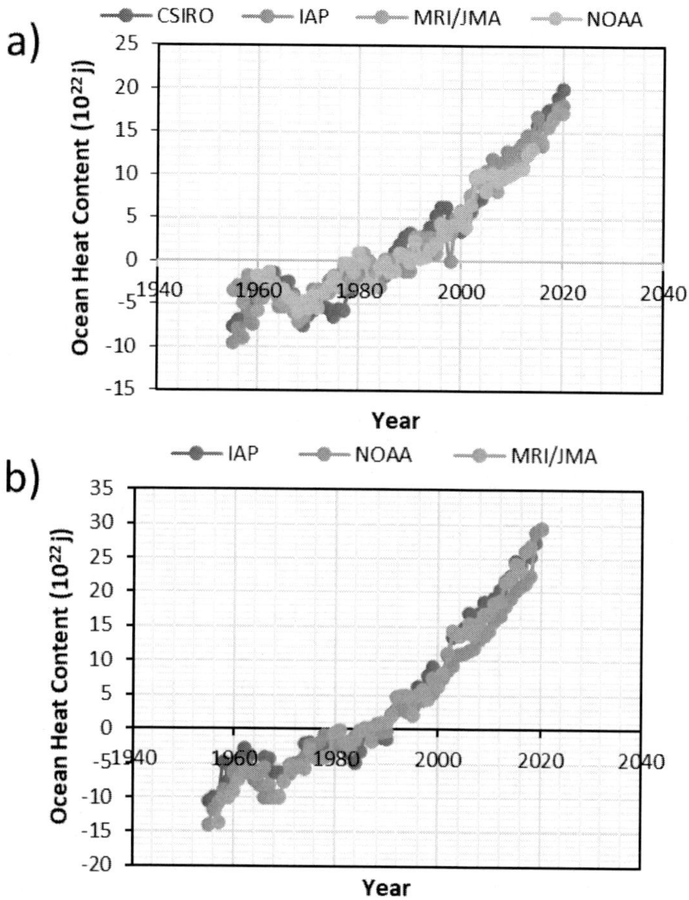

Figure 8. Ocean heat content at (a) 100 and (b) 2000 meters of the oceans (drawn by the authors).

From our analysis, a clear evolutionary trajectory of research concerning climate change and molluscs between 1995 and 2015 emerges. In the 1990s, the historical data accentuated the adaptability and resilience of the fisheries and aquaculture sector to changes, signaling an industry well-acquainted with shifts and challenges. As the new millennium dawned, there was a palpable shift in research emphasis towards understanding the implications of water acidification on marine ecosystems (Ishii et al., 2017; Cheng et al., 2022). This trend gained momentum and remained dominant up to the early 2010s, with research elucidating the potentially adverse effects of ocean acidification on a multitude of marine and estuarine molluscs (Mitchell et al., 2010; Guinotte et al., 2008).

Our findings underscore that ocean acidification became a focal point for researchers during this period. Complementing this, there was also an increased interest in examining the temporal and spatial variations of climate change impacts, as evidenced by the prominence of research clusters centered on seasonal and geographical variations. Given the global implications of climate change, this broad-scale and longitudinal focus in research is both intuitive and imperative. Understanding these patterns at vast scales and over extended periods is essential for grasping the multifaceted and ever-evolving dynamics of our planet's changing climate.

Figure 9 presents global CO2 emissions from 1990 to 2015, as documented by Morningstar Sustainalytics (2022) and FAOSTAT (2022). The data indicates a noteworthy 51% surge in net CO2 emissions during this period. This rise is significant, as it accounts for nearly three-quarters of total global emissions (US EPA, 2022). As illustrated in Figure 9, this trend aligns with findings from various studies monitoring atmospheric CO2 concentrations (Bereiter et al., 2015; Global Monitoring Laboratory, 2022; CSIRO GASLAB, 2022; Flask Measurements at Lampedusa Island, 2022). There has been a pronounced escalation in CO2 concentrations since the onset of the industrial era. Specifically, from the late 1700s annual average of 280 ppm, levels rose to 414 ppm in 2021 – an increase of 48%. This augmentation is predominantly attributed to human activities (US EPA, 2022).

Measurements over the last few decades have revealed that rising levels of ocean CO_2 are a result of rising levels of atmospheric CO_2 [40]. The acidity rose because of this circumstance (i.e., a fall in pH). Historical models (Figure 9) show that since the 1880s, increasing CO_2 levels have led to a decline in the saturation level of aragonite in oceans all over the world, making it more difficult for molluscs to build and maintain their shells. In fact, as seawater's acidity rises, its ionic equilibrium becomes unstable, and it becomes more hostile toward $CaCO_3$ (Eq. 6) (Nunes, 2023).

$$CaCO_3 + CO_2 + H_2O \square Ca(HCO_3)_2 \qquad (6)$$

However, most studies have since the 2010s concentrated on oxidative stress, genetic traits, and the impact of climate change on aquatic communities and their relative adaptive response. (US EPA, 2022; FAOSTAT, 2022; FAO 2020).

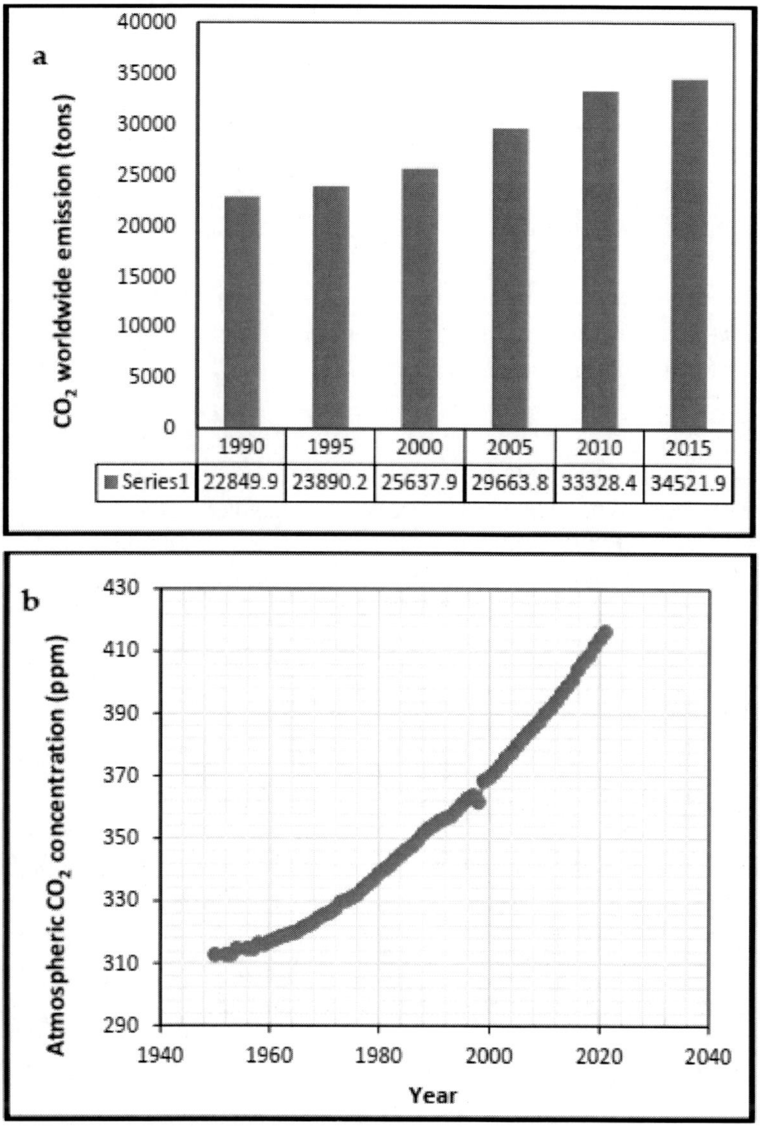

Figure 9. Greenhouse gas concentrations in the atmosphere (a) The CO_2 worldwide emission from 1990 to 2015 (b) CO_2 concentrations in the atmosphere between 1950 and 2021 (drawn by the authors).

It is very difficult, or almost impossible to predict future trends on climate change and molluscs group. Nonetheless, recent studies on the climate change are almost following the same trend as the last 20 years, specifically concerning acidification (Mejia et al, 2021) and it looks like it's going to last, at least for the next few years.

By quantitatively analyzing our results, we can attest that in the study domain of climate change associated with the molluscs group, the publications progressed very slowly, and increased rapidly from 2011 until 2021. The increasing number of editions available in the WOS may explain this. For example, further research should identify the difference between WOS editions within the Core Collection.

Based on our latest findings, collections within the database can be segmented into four distinct categories based on their years of inception, as of 1st October 2022. These categories are:

1. Science Citation Index Expanded and Social Sciences Citation Index (SSCI) span from 1975 to the present day. Similarly, Arts & Humanities Citation Index (AHCI) has been active from 1975 and continues presently.
2. Both Conference Proceedings Citation Index – Science (CPCI-S) and Conference Proceedings Citation Index – Social Science & Humanities (CPCI-SSH) cover a timeline from 1990 up to the present.
3. The Book Citation Index - Science (BKCI-S) and Book Citation Index – Social Sciences & Humanities (BKCI–SSH) started their collections from 2005 and persist to the present day.
4. The most recent addition, Emerging Sources Citation Index (ESCI), began in 2017 and continues up to the present. For a comprehensive scientometric review, data from 2017 onward should be particularly considered.

Twenty distinct co-citation clusters have been identified from the analysis. The cluster such as "ocean acidification," "seasonal variation," "whelk busycon-contrarium (Conrad)" is among the top three clusters in the field. Previous studies also showed that a few commercial aquaculture species in the world were also impacted by the ocean acidification and are being linked with climate change (Gazeau et al., 2007; Gazeau et al., 2013). The evolution of the cited document also suggested that the article published by Hawkins et al. (Hawkins et al., 1986) is the most central document in the field. It is not surprising that the most influential authors and journals are mostly researchers and titles from the twentieth century, because the influences have been accumulated to date; however, most of the technologies/methodologies used at the time have already been replaced by a more advanced technology today, yet the old papers are still being cited as references.

The bibliometric analysis (Table 1) of the research related to molluscs and climate change showcases the academic community's evolving interests and the significance of certain seminal works:

- Temporal Distribution of High-Impact Publications: The analysis spotlights the most impactful studies primarily occurring between 2012 and 2015. This concentration suggests a period of particularly innovative or pioneering work in this field. These years could have witnessed breakthroughs or important findings that shaped subsequent research, indicated by the high number of citations these works garnered.
- Lag in High-Impact Publications: The absence of high-impact papers from 2016 to 2022 might not indicate a lack of valuable research during these years. Instead, as the analysis pointed out, citations from recent years might not have accumulated sufficiently to outpace the impact of prior work. As newer studies mature and get incorporated into subsequent research, their citation count may rise, potentially altering the bibliometric landscape in the future.
- Geographical Distribution: The significant representation of the USA, with main authors from five of the top ten publications being based there, aligns with the country's history of robust research infrastructure, funding, and a strong academic community. China's growing prominence in global research is also reflected in this data, most notably in the form of Guofan Zhang's monumental study from the Institute of Oceanology, Chinese Academy of Science. This work's remarkable citation strength showcases its foundational significance in the field.
- Global Landscape: The broader implication, as observed by Wagner et al., 2022, about the dominance of China and the USA in global research is reflected in this analysis. Their research infrastructure, investments, and academic environments have rendered them as central players in the global academic arena.

5. Limitations

Considering the constraint that the analytical tool can integrate only one database in a single run, it's advised to consider incorporating other expansive article databases in future analyses. Databases like SCOPUS, Pub-Med, China National Knowledge Infrastructure (CNKI), Chinese Social Sciences Citation Index (CSSCI), and Index Copernicus offer in-depth details on articles, including titles, authors, abstracts, citations, and co-citations. Incorporating these databases can enhance the breadth and depth of the research analysis.

6. Future Researches Direction

Certainly, a deeper dive into this topic is imperative to gain a comprehensive understanding of the impact of climate change on molluscs, highlighting the pressing need to alert relevant authorities about combatting climate change, curbing pollution, and rejuvenating degraded ecosystems. Subsequent scientometric analyses could delve into:

- Temporal Evolution in Non-Scientific Publications: Exploring how non-scientific publications, sourced from databases like SSCI and A&HCI, have evolved over time in addressing climate change, global warming, and their implications on marine ecosystems. Molluscs, among other taxa, can be considered as exemplars.
- Comparative Analysis of Climate Change Literature: Drawing contrasts between scientific papers discussing the causatives and ramifications of climate change, versus those proposing mitigation or adaptation strategies.
- Spotting Literature Gaps: Recognizing prominent thematic voids within the literature. This might encompass overlooking highly cited articles, or content from journals boasting significant impact factors, as well as papers published post-2021.

Furthermore, the findings from this research can be harnessed to pave the way for refined research strategies centered around molluscs, acknowledging both their ecological significance and economic value.

Conclusion

The present study shows that the links between climate change and molluscs group are vital for a better understanding of the research direction in this domain. As climate change continues to disrupt weather patterns, researchers around the globe are focusing on strategies to mitigate its impact, including reducing the risk of disasters, managing natural resources, and addressing issues such as water warming, acidification, and species adaptation. The measurable effects of climate change are already evident in various natural systems, particularly among marine organisms. Therefore, climate change remains one of the foremost concerns for researchers worldwide. The importance of studying the consequences of climate change and molluscs for sustainable development is highlighted by the fact that climate change has been one of the most extensively studied subjects. However, further collaboration and research are required to fully address the concerns associated with this field.

The current study provides a comprehensive analysis and summary of the current trends and information on publication output, countries, authors, hot topics, and impactful keywords in the field of climate change and molluscs. Furthermore, current research suggests that climate change may alter molluscs in the next decades. The upsurge of the number of publications of study related to climate change and molluscs in the last 10 years indicate the importance of this shelled organism as a good bioindicator for climate impacts on the ecosystem. Interesting data generated from the manuscript have been shared and are available to be compared and used by other researchers. Undoubtedly, more work needs to be done conducted in the field to identify any significant impacts of climate driven changes on coastal environments and specifically on molluscs in view of their ecological and economic importance.

Author Contributions

Conceptualization, M. N. A. and R. B. A. L.; methodology, M. I. M. N and W. M. R. W. H.; software, R. B. A. L and M. N. A.; validation, W. M. R. W. H.; formal analysis, M. N. A and M. I. M. N.; investigation, R. B. A. L.; resources M. N. A.; data curation, R. B. A. L and W. M. R. W. H.; writing—original draft preparation, M. N. A; writing—review and editing, M. N. A and S. B.; visualization, S. B.; supervision, W. M. R. W. H; project administration,

M. N. A.; funding acquisition, M. N. A.; editing, M. N. I. K. All authors agreed to the published version of the manuscript.

Funding

The coloured figures for the present chapter has been sponsored with funds provided by USDA-NIFA Sustainable Agriculture Systems, Grant No. 2019-69012-29905 to one of the Editors, Prof. Dr. Guillermo Téllez-Isaías. Title of the project: Empowering US Broiler Production for Transformation and Sustainability USDA-NIFA (Sustainable Agriculture Systems): No. 2019-69012-29905.

Data Availability Statement

Data supporting reported results are available from the corresponding author.

Acknowledgments

The present work was supported by the Long-Term Research Grant Scheme (LRGS) awarded by the Department of Higher Education, Ministry of Higher Education Malaysia (LRGS/1/2020/UMT/01/1; LRGS UMT Vot No. 56040) (2020—2024).

Conflicts of Interest

The authors declare no conflict of interest.

References

Ablah, E. Olson, C. Brown, J. Lee, C. A. Konda, K., Bronleewe, T. Buzard, A. 2018. Need for Transparency and Ongoing Communication After Residents With Contaminated Water Wells Are Connected to City Water. *Journal of Environmental Health.* 81, 26–31.

Atmospheric Carbon Dioxide Record from Flask Measurements at Lampedusa Island Available online: https://cdiac.ess-dive.lbl.gov/trends/co2/lampis.html (accessed on 10 December 2022).

Atmospheric CO2 Concentrations from the CSIRO GASLAB Flask Sampling Network Available online: https://cdiac.ess-dive.lbl.gov/trends/co2/csiro/ (accessed on 10 December 2022).

Azra, M. N. Noor, M. I. M. Sung, Y. Y. Lawrence, E. R. Ghaffar, M. A. 2022. What Evidence Exists on the Impact of Climate Change on Some of the Worst Invasive Fish and Shellfish? A Systematic Map Protocol. *Environmental Evidence.* 11, 19, doi:10.1186/s13750-022-00273-z.

Bereiter, B. Eggleston, S. Schmitt, J. Nehrbass-Ahles, C. Stocker, T. F. Fischer, H. Kipfstuhl, S., Chappellaz, J. 2015. Revision of the EPICA Dome C CO 2 Record from 800 to 600 Kyr before Present: Analytical Bias in the EDC CO2 Record. *Geophysical Research Letters*, 42, 542–549, doi:10.1002/2014GL061957.

Boroda, A. V., Kipryushina, Y. O., Odintsova, N. A. 2020. The Effects of Cold Stress on Mytilus Species in the Natural Environment. *Cell Stress Chaperones*, 25, 821–832, doi:10.1007/s12192-020-01109-w.

Boyer, T. *Global Ocean Heat and Salt Content* - Seasonal, Yearly, and Pentadal Fields Available online: https://www.ncei.noaa.gov/access/global-ocean-heat-content/ (accessed on 23 December 2022).

Brierley, A. S., Kingsford, M. J. (2009). Impacts of climate change on marine organisms and ecosystems. *Current biology*, 19(14), R602-R614. https://doi.org/10.1016/j.cub.2009.05.046.

Byrne, M. 2011. *Impact of Ocean Warming and Ocean Acidication on Marine Invertebrate Life History Stages*: Vulnerabilities and Potential for Persistence in a Changing Ocean. In *Oceanography and Marine Biology*; CRC Press, ISBN 978-0-429-10992-8.

Carbon Emissions Data Available online: https://www.sustainalytics.com/investor-solutions/esg-research/climate-solutions/carbon-emissions-data (accessed on 10 December 2022).

Chen, C. 2020. *How to Use CiteSpace*. LeanPub Publishing.

Cheng, L.; Abraham, J.; Trenberth, K. E.; Fasullo, J.; Boyer, T.; Mann, M. E.; Zhu, J.; Wang, F.; Locarnini, R.; Li, Y. Reseghetti, F., Simoncelli, S., Gouretski, V., Chen, G., Mishonov, A., Reagen, J. 2022. Another Record: Ocean Warming Continues through 2021 despite La Niña Conditions. *Advances in Atmospheric Sciences*. 39, 373–385, doi:10.1007/s00376-022-1461-3.

Climate Change 2007: *The Physical Science Basis: Contribution of Working Group I* to the Fourth Assessment Report of the Intergovernmental Panel on Climate Change; Solomon, S., Intergovernmental Panel on Climate Change, Intergovernmental Panel on Climate Change, Eds.; Cambridge University Press: Cambridge; New York, 2007; ISBN 978-0-521-88009-1.

Climate Change: *Science and Solutions for Australia*; CSIRO, 2006; ISBN 978-0-643-10326-9.

Doney, S. C. Fabry, V. J. Feely, R. A. Kleypas, J. A. 2009. Ocean Acidification: The Other CO2 Problem. *Annual Review of Marine Science*. 1, 169–192, doi:10.1146/annurev.marine.010908.163834.

Fabry, V. J. Seibel, B. A. Feely, R. A. Orr, J. C. 2008. Impacts of Ocean Acidification on Marine Fauna and Ecosystem Processes. ICES *Journal of Marine Science*. 65, 414–432, doi:10.1093/icesjms/fsn048.

FAOSTAT. 2022. *Food and Agriculture Statistics*. Available online: https://www.fao.org/faostat/en/#data/EL (accessed on 10 December 2022).

Fortunato, S., Hric, D. 2016. Community Detection in Networks: A User Guide. Physics Reports. 659, 1–44, doi:10.1016/j.physrep.2016.09.002.

Gazeau, F., Parker, L. M., Comeau, S., Gattuso, J. P., O'Connor, W. A., Martin, S. Pörtner, H. O., Ross, P. M. 2013. Impacts of Ocean Acidification on Marine Shelled Molluscs. *Marine Biology*. 160, 2207–2245, doi:10.1007/s00227-013-2219-3.

Gazeau, F., Quiblier, C., Jansen, J. M., Gattuso, J. P., Middelburg, J. J., Heip, C. H. R. 2007. Impact of Elevated CO2 on Shellfish Calcification. *Geophysical Research Letters*. 34, doi:10.1029/2006GL028554.

Gordon, T. A. C., Wilding, E. L., Aldridge, D. C. 2016. Predation of Freshwater Gastropods (*Viviparus viviparus*) by Brown Rats (Rattus Norvegicus). *Journal of Molluscan Studies*. 82, 457–463, doi:10.1093/mollus/eyw012.

Guinotte, J. M., Fabry, V. J. 2008. Ocean Acidification and Its Potential Effects on Marine Ecosystems. *Annals of the New York Academy of Sciences*. 1134, 320–342, doi:10.1196/annals.1439.013.

Hale, R., Calosi, P., McNeill, L., Mieszkowska, N., Widdicombe, S. 2011. Predicted Levels of Future Ocean Acidification and Temperature Rise Could Alter Community Structure and Biodiversity in Marine Benthic Communities. *Oikos*. 120, 661–674, doi:10.1111/j.1600-0706.2010.19469.x.

Hawkins, A. J. S., Bayne, B. L., Day, A. J., Denton, E. J.. 1986. Protein Turnover, Physiological Energetics and Heterozygosity in the Blue Mussel, Mytilus Edulis: The Basis of Variable Age-Specific Growth. *Proceedings of the Royal Society B: Biological Sciences*. 229, 161–176, doi:10.1098/rspb.1986.0080.

Hendriks, I. E., Duarte, C. M., Álvarez, M. 2010. Vulnerability of Marine Biodiversity to Ocean Acidification: A Meta-Analysis. *Estuarine, Coastal and Shelf Science*. 86, 157–164, doi:10.1016/j.ecss.2009.11.022.

Hou, D., Bi, X., Mao, Z., Fan, Y., Hu, X., Li, X. 2019. Biomaterials Research of China from 2013 to 2017 Based on Bibliometrics and Visualization Analysis. *PeerJ*. 7, e6859, doi:10.7717/peerj.6859.

Ishii, M., Fukuda, Y., Hirahara, S., Yasui, S., Suzuki, T., Sato, K. 2017. Accuracy of Global Upper Ocean Heat Content Estimation Expected from Present Observational Data Sets. *SOLA*. 13, 163–167, doi:10.2151/sola.2017-030.

Kroeker, K. J. Kordas, R. L. Crim, R. Hendriks, I. E. Ramajo, L. Singh, G. S. Duarte, C. M. Gattuso, J. P. 2013. Impacts of Ocean Acidification on Marine Organisms: Quantifying Sensitivities and Interaction with Warming. *Global Change Biology*. 19, 1884–1896, doi:10.1111/gcb.12179.

Kroeker, K. J. Kordas, R. L. Crim, R. N., Singh, G. G. 2010. Meta-Analysis Reveals Negative yet Variable Effects of Ocean Acidification on Marine Organisms. *Ecological Letter*. 13, 1419–1434, doi:10.1111/j.1461-0248.2010.01518.x.

Li, Q. Sun, S., Zhang, F., Wang, M., Li, M. 2019. Effects of Hypoxia on Survival, Behavior, Metabolism and Cellular Damage of Manila Clam (Ruditapes Philippinarum). *PLOS ONE*. 14, e0215158, doi:10.1371/journal.pone.0215158.

Mejia, C., Wu, M., Zhang, Y. 2021. Kajikawa, Y. Exploring Topics in Bibliometric Research. Through Citation Networks and Semantic Analysis. *Frontiers in Research Metrics and Analytics*. 6, 742311, doi:10.3389/frma.2021.742311.

Miao, Y., Zhang, Y., Yin, L. 2018. Trends in Hepatocellular Carcinoma Research from 2008 to 2017: A Bibliometric Analysis. *PeerJ*, 6, e5477, doi:10.7717/peerj.5477.

Minasny, B., Hartemink, A. E., McBratney, A. Jang, H. J. 2013. Citations and the h Index of Soil Researchers and Journals in the Web of Science, Scopus, and Google Scholar. *PeerJ*, 1, e183, doi:10.7717/peerj.183.

Mitchell, M. J. Jensen, O. E. Cliffe, K. A. Maroto-Valer, M. M. 2010. A Model of Carbon Dioxide Dissolution and Mineral Carbonation Kinetics. *Proceedings of the Royal Society of London. Series A, Mathematical and Physical Sciences.* 466, 1265–1290, doi:10.1098/rspa.2009.0349.

Moras, C. A. Bach, L. T. Cyronak, T. Joannes-Boyau, R. Schulz, K. G. 2022. Ocean Alkalinity Enhancement – Avoiding Runaway CaCO3 Precipitation during Quick and Hydrated Lime Dissolution. *Biogeosciences.* 19, 3537–3557, doi:10.5194/bg-19-3537-2022.

National Oceanic and Atmospheric Administration. (2021, April 6). *Carbon dioxide now more than 50% higher than pre-industrial levels.* Retrieved from https://www.noaa.gov/news-release/carbon-dioxide-now-more-than-50-higher-than-pre-industrial-levels.

Nunes, L. J. R. 2023. The Rising Threat of Atmospheric CO2: A Review on the Causes, Impacts, and Mitigation Strategies. *Environments.* 10, 66. https://doi.org/10.3390/environments10040066.

Parisi, V., Gandolfi, G. 1974. Further Aspects of the Predation by Rats on Various Mollusc Species. *Bollettino della Società zoologica italiana.* 41, 87–106, doi:10.1080/11250007409430096.

Prentice, R. L. 2011 - *Statistical Methods and Challenges in Epidemiology and Biomedical Research*. In *Essential Statistical Methods for Medical Statistics*; Rao, C. R., Miller, J. P., Rao, D. C., Eds.; North-Holland: Boston, pp. 1–26 ISBN 978-0-444-53737-9.

Regoli, F. Giuliani, M. E. 2014. Oxidative Pathways of Chemical Toxicity and Oxidative Stress Biomarkers in Marine Organisms. *Marine Environmental Research.* 93, 106–117, doi:10.1016/j.marenvres.2013.07.006.

Ries, J. B., Cohen, A. L., McCorkle, D. C. 2009. Marine Calcifiers Exhibit Mixed Responses to CO2-Induced Ocean Acidification. *Geology* 37, 1131–1134, doi:10.1130/G30210A.1.

Schöne, B. R. 2013. Arctica Islandica (Bivalvia): A Unique Paleoenvironmental Archive of the Northern North Atlantic Ocean. *Global Planetary Change*, 111, 199–225, doi:10.1016/j.gloplacha.2013.09.013.

Sowa, A., Krodkiewska, M., Halabowski, D., Lewin, I. 2019. Response of the mollusc communities to environmental factors along an anthropogenic salinity gradient. *Science Nature.* 106, 60 https://doi.org/10.1007/s00114-019-1655-4.

Tan, K. Zheng, H. 2020. Ocean Acidification and Adaptive Bivalve Farming. *Science of Total Environment.* 701, 134794, doi:10.1016/j.scitotenv.2019.134794.

Thirukanthan, C. S., Azra, M. N., Seman, N. J. A., Agos, S. M., Arifin, H., Aouissi, H. A., Gao, H. (2023). A scientometric review of climate change and research on crabs. *Journal of Sea Research*, 193, 102386. https://doi.org/10.1016/j.seares.2023.102386.

Thomsen, J. Casties, I. Pansch, C. Körtzinger, A. Melzner, F. 2013. Food Availability Outweighs Ocean Acidification Effects in Juvenile Mytilus Edulis: Laboratory and Field Experiments. *Global Change Biology*. 19, 1017–1027, doi:10.1111/gcb.12109.

Ullah Khan, F., Shang, Y., Chang, X. Kong, H. Zuberi, A. Fang, J. K. H. Liu, W. Peng, J. Zhang, X. Hu, M., Wang, Y. 2021. Effects of Ocean Acidification, Hypoxia, and Warming on the Gut Microbiota of the Thick Shell Mussel Mytilus Coruscus Through 16S RRNA Gene Sequencing. *Frontier in Marine Science*., 8. 736338.

US Department of Commerce, N. *Global Monitoring Laboratory* - Carbon Cycle Greenhouse Gases Available online: https://gml.noaa.gov/ccgg/trends/data.html (accessed on 10 December 2022).

US EPA, O. *Climate Change Indicators*: Atmospheric Concentrations of Greenhouse Gases Available online: https://www.epa.gov/climate-indicators/climate-change-indicators-atmospheric-concentrations-greenhouse-gases (accessed on 23 December 2022).

US EPA, O. *Climate Change Indicators*: Global Greenhouse Gas Emissions Available online: https://www.epa.gov/climate-indicators/climate-change-indicators-global-greenhouse-gas-emissions (accessed on 10 December 2022).

US EPA, O. *Climate Change Indicators*: Sea Surface Temperature Available online: https://www.epa.gov/climate-indicators/climate-change-indicators-sea-surface-temperature (accessed on 23 December 2022).

Wagner, C. S. Zhang, L. Leydesdorff, L. 2022. A Discussion of Measuring the Top-1% Most-Highly Cited Publications: Quality and Impact of Chinese Papers. *Scientometrics.* 127, 1825–1839, doi:10.1007/s11192-022-04291-z.

Waldbusser, G. G. Hales, B. Langdon, C. J. Haley, B. A. Schrader, P., Brunner, E. L. Gray, M. W. Miller, C. A. Gimenez, I. 2015. Saturation-State Sensitivity of Marine Bivalve Larvae to Ocean Acidification. *Nature Climate Change*. 5, 273–280, doi:10.1038/nclimate2479.

Wang, J. Ren, R. Yao, C. L. 2018. Oxidative Stress Responses of Mytilus Galloprovincialis to Acute Cold and Heat during Air Exposure. J. *Molluscan Studies*. 84, 285–292, doi:10.1093/mollus/eyy027.

Wittmann, K. J. 2013. Comparative Morphology of the External Male Genitalia in Lophogastrida, Stygiomysida, and Mysida (Crustacea, Eumalacostraca). *Zoomorphology*. 132, 389–401, doi:10.1007/s00435-013-0198-z.

Zhang, G., Fang, X., Guo, X., Li, L., Luo, R., Xu, F., Yang, P., Zhang, L., Wang, X., Qi, H., Xiong, Z., Que, H., Xie, Y., Holland, P. W., Paps, J., Zhu, Y., Wu, F., Chen, Y., Wang, J., Peng, C., Meng, J., Yang, L., Liu, J., Wen, B., Zhang, N., Huang, Z., Zhu, Q., Feng, Y., Mount, A., Hedgecock, D., Xu, Z., Liu, Y., Domazet-Lošo, T., Du, Y., Sun, X., Zhang, S., Liu, B., Cheng, P., Jiang, X., Li, J., Fan, D., Wang, W., Fu, W., Wang, T., Wang, B., Zhang, J., Peng, Z., Li, Y., Li, N., Wang, J., Chen, M., He, Y., Tan, F., Song, X., Zheng, Q., Huang, R., Yang, H., Du, X., Chen, L., Yang, M., Gaffney, P. M., Wang, S., Luo, L., She, Z., Ming, Y., Huang, W., Zhang, S., Huang,

B., Zhang, Y., Qu, T., Ni, P., Miao, G., Wang, J., Wang, Q., Steinberg, C. E., Wang, H., Li, N., Qian, L., Zhang, G., Li, Y., Yang, H., Liu, X., Wang, J., Yin, Y., Wang. J. 2012. The Oyster Genome Reveals Stress Adaptation and Complexity of Shell Formation. *Nature*, 490, 49–54, doi:10.1038/nature11413.

Chapter 5

Scientometrics of Climate Change and Shrimp Diseases: An Overview

Kamariah Bakar[1,*]
Mohd Ihwan Zakariah[2,†]
Mohd Iqbal Mohd Noor[3,4]
Mazlan Abd Ghaffar[1,2,5]
Mohamad Nor Azra[1,6]
Thirukanthan Chandra Segaran[1]
Hidir Ariffin[2]
Guillermo Téllez-Isaías[7]
Zulhisyam Abdul Kari[8]
and Walter G. Bottje[7]

[1]Institute of Climate Adaptation and Marine Biotechnology (ICAMB), University Malaysia Terengganu (UMT), Terengganu, Malaysia
[2]Institution Centre of Excellence (HICoE), Institute of Tropical Aquaculture and Fisheries (AKUATROP), University Malaysia Terengganu, Terengganu, Malaysia
[3]Faculty of Business Management, University Technology MARA (UiTM) (Pahang), Raub, Pahang, Malaysia
[4]Institute for Biodiversity and Sustainable Development, University Technology MARA (UiTM), Shah Alam, Selangor, Malaysia
[5]Faculty of Science and Marine Environment, University Malaysia Terengganu, Terengganu, Malaysia
[6]Research Center for Marine and Land Bioindustry, Earth Sciences and Maritime Organization, National Research and Innovation Agency (BRIN), Pemenang, West Nusa Tenggara, Indonesia

---

[*] Corresponding Author's Email: kamariah@umt.edu.my.
[†] Corresponding Author's Email: ihwanz@umt.edu.my.

In: Marine Life in Changing Climates
Editors: M. Nor Azra, T. Chandra Segaran, G. Téllez-Isaías et al.
ISBN: 979-8-89113-404-1

[7]Department of Poultry Science, University of Arkansas, Fayetteville, USA
[8]Department of Agricultural Science, Faculty of Agro-Based Industry, University Malaysia Kelantan, Jeli, Kelantan, Malaysia

Abstract

According to FAO fisheries statistics, global shrimp production has shown an upward trend. Shrimp stand as an essential aspect of coastal fisheries for many countries worldwide. With climatic factors like fluctuations in sea surface temperature, elevating sea levels, coastal inundations, eroding coastlines, oceanic heatwaves, and heightened ocean acidity, there's growing concern for regions spanning both coastal and open oceans. This is especially significant as these are the primary zones for a majority of aquaculture operations. This chapter aims to synthesize the available literature on the impacts of climate change on global shrimp disease using the Scientometric method. We first identified the top disease in shrimp aquaculture along with climate change elements, then extracted bibliometrics from the Web of Science. Then we used CiteSpace to assess trends and research focusing on the impacts of climate change on worldwide shrimp disease. To address gaps in this emerging field, we make two main recommendations: (i) Increased collaboration between countries to develop global solutions towards shrimp disease* in the face of climate change and (ii) Drawing from other disciplines outside of "Ecology, Earth, Marine," such as molecular biology, economics, and sustainability, would strengthen and provide additional insights on sustainable shrimp production in the world. Further qualitative discussion was provided along with the future research topic's direction of Sustainable Development Goals of Food Security and Climate Change.

Keywords: aquaculture, environmental science, global warming, white spot syndrome virus, white leg shrimp, climate change, shellfish

1. Introduction

Shellfish cultivation stands as a pivotal sector within global captive fisheries and aquaculture (Ramli et al. 2022; Azra et al. 2021). It's essential to regularly update the list of commercial shellfish species identified as potential candidates for aquaculture to ensure resource sustainability and security. Data from the Food and Agriculture Organization (FAO) indicate that shellfish have become a primary aquaculture product, increasingly serving as a key protein source. However, challenges persist that demand immediate attention to uphold the shellfish aquaculture industry's vital role in meeting global food security needs (Boyd et al. 2022; Rocha et al. 2022). One such challenge is the prevalence of diseases in shellfish farming, which poses significant risks to food production stability (Brooks et al. 2022; Stentiford et al. 2012). Diseases, indeed, rank among the gravest threats facing aquaculture production (Doan et al. 2022; El-Saadony et al. 2022). There's an urgent need for rigorous diagnosis of parasites and consistent monitoring for disease pathogens in both farmed and wild shellfish populations. Such measures are critical to safeguarding the future of the global aquaculture sector. They will also ensure the longevity and health of pristine shellfish groups, bolstering the overall industry. Shrimp, a type of shellfish, has emerged as one of the most widely farmed crustaceans globally, attributed to its rapid growth, prolific yield, and adaptability to varying temperature and salinity conditions, even under intensive farming setups.

Similar to other aquaculture species, shrimp is also exposed to the impacts of global climate change, which include the increased occurrence of extreme weather events resulting in drastic fluctuations to abiotic factors such as the levels of oceanic pH, carbon dioxide, dissolved oxygen (DO), water temperature and salinities (Galappathhi et al., 2020). The rapid alterations to these abiotic factors may result in variations in viral and bacterial loads, causing severe physiological stress to cultured species leading to disease outbreaks, and devastating economic losses in the shrimp farming industry.

Adopted in 2015, the Paris Agreement sets forth an ambitious objective: to amplify the global reaction to climate change by ensuring that this century's global temperature increase remains considerably below 2 degrees Celsius compared to pre-industrial times. Additionally, the accord strives to fortify countries' resilience against climate change ramifications. This is achieved by promoting adequate financial allocations, introducing a novel technological structure, and bolstering capacity-building frameworks. Given current concentrations and on-going emissions of greenhouse gases, it is likely that by the end of this century, the increase in global temperature will exceed 1.5°C compared to 1850 to 1900 for all but one scenario. The world's oceans will warm and ice melt will continue. Average sea level rise is predicted as 24 – 30 cm by 2065 and 40-63 cm by 2100. Most aspects of climate change will persist for many centuries even if emissions are stopped (Dimitrov, 2016).

Concurrent global climate change patterns strongly suggest a heightened risk of disease outbreaks in shellfish aquaculture (Maulu et al. 2021; Rowley et al. 2014). Adverse impacts on fisheries and aquaculture, stemming from global warming, human-induced alterations, and contamination, can precipitate production losses (Azra et al. 2022; Braña et al. 2021), potentially paving the way for future infectious outbreaks. Such infections in economically significant shellfish could facilitate the transfer of certain disease agents, possibly shaping the prevalence and severity of shellfish diseases in the times to come. This study offers invaluable foundational insights into the trajectory and evolution of literature examining the influence of climate change factors on shrimp diseases globally. The increased accessibility of citation data is commendable, ushering in an era of enhanced transparency and accountability for research efforts. This, in turn, aids scholars, benefactors, and policymakers in making more informed decisions regarding focal research areas.

Human activities are accelerating climate change at an unparalleled pace, leading to swift alterations in abiotic factors within aquatic ecosystems. The aquaculture sector is a significant contributor to carbon dioxide (CO_2) emissions, further elevating the levels of greenhouse gases (GHG) in the atmosphere. CO_2 plays a major role in global warming by absorbing and subsequently re-emitting infrared light. A surplus of CO_2 emissions in the atmosphere is also assimilated by the oceans, altering their buffering capacity and overall chemistry. This process, in turn, results in the lowering of seawater pH. Rapid changes of any or combination of abiotic factors such as temperature, salinity, pH, total ammonia (TAN) and dissolved oxygen (DO) could affect the biology, physiology, growth and population dynamics of aquatic species. While the widely cultured Pacific White Shrimp, *Litopenaeus vannamei*, boasts a broad adaptability to variations in salinity and temperature, acute shifts in these abiotic factors, especially water acidification, can pose significant threats. Persistent atmospheric CO_2 emissions in densely populated ponds can swiftly decrease pH levels, potentially causing DNA damage, impairing immunity, compromising osmoregulation capabilities, and heightening disease vulnerability in shrimp. This may culminate in substantial mortality rates, resulting in major socioeconomic setbacks for the shrimp industry. Repeated disease outbreaks and stunted growth, stemming from pronounced pH fluctuations and other abiotic stressors, could undermine the financial viability, wellbeing, and societal stature of communities engaged in shrimp cultivation over extended periods. Therefore, enhancing our understanding of the interplay between climate change, shrimp diseases, and strategies to mitigate global warming's effects on aquatic chemistry and the planet at large is paramount (Flegel et al. 2008).

Shrimp represents a vital element of coastal fisheries resources in numerous countries, with Asia being especially notable in this context. Factors such as sea surface temperature, rising sea levels, coastal flooding, coastal erosion, oceanic heatwaves, and increased ocean acidity comprise several climate-related drivers anticipated to influence both coastal and open ocean regions. This becomes particularly significant given that these regions predominantly host a majority of aquaculture operations.

Thus, the main aim of this study is to synthesize the available literature on the impacts of climate change on shrimp disease in Asia using the Scientometric method. The specific objectives were to identify its trends through the (i) evolution of publications (number of articles, authors involved, contributing affiliations and journals), (ii) dual map overlay, (iii) citation analysis of Author, Journal & document and (iv) cluster network.

2. Methods

2.1. Data Sources

We conducted a search for pertinent English-language research articles within the Web of Science Core Collection (WOSCC) database. Our search strategy employed two primary keywords: "climate change" and "shrimp disease." In the WOSCC, the search was executed using the "topic" (TS) field, which encompasses titles, abstracts, keywords, and "Keywords Plus" from cited articles. We considered articles from the inception of the database up until the 30th of June 2022, the date on which our search was performed. The specific keywords applied in this study were "climate change" and "shrimp disease." Their synonyms are delineated below:

2.1.1. Climate Change

TS=(("climat*") OR ("climat* chang*") OR ("anthropogenic") OR ("global warm*") OR ("seasonal* variat*") OR ("extrem* event*") OR ("environment* variab*") OR ("anthropogenic effect*") OR ("greenhouse") OR ("greenhouse effect*") OR ("sea level ris*") OR ("sea level") OR (erosio*) OR ("agricult* runof") OR ("weather* variab*") OR ("weather* extrem*") OR ("extreme* climat*") OR ("environment* impact*") OR ("environment* chang*") OR ("anthropogenic stres*") OR ("temperature ris*") OR ("temperature efect*") OR ("warm* ocean") OR ("sea surface* temperat*") OR (heatwav*) OR (acidi*) OR ("acidification") OR (hurrican*) OR (el nino) OR ("el-nino") OR ("la nina") OR (la-nina) OR (drought*) OR (flood*) OR ("high precipit*") OR ("heavy rainfall*") OR ("CO_2 concentrat*") OR ("melt* of the glacier*") OR ("melt* ice*") OR ("therm* stress*") OR (hypoxia) OR ("mean surface temperature")).

2.1.2. Shrimp Disease

TS = ((shrimp diseas*) OR ("White spot syndrome") OR ("White spot syndrome virus") OR ("Infectious hypodermal and hematopoeitic necrosis virus") OR ("Hepatopancreatic parvovirus") OR ("Spawner-isolated mortality virus") OR ("Lymphoidal parvo-like virus") OR ("Baculovirus penaei") OR ("Monodon baculovirus") OR ("Baculovirus midgut gland necrosis virus") OR ("Type C baculovirus of Penaeus monodon") OR ("Shrimp iridovirus") OR ("Shrimp hemocyte iridescent virus") OR ("Taura syndrome virus") OR ("Infectious myonecrosis virus") OR ("Covert mortality nodavirus") OR ("Penaeus vannamei nodavirus") OR ("Yellow head virus") OR ("Gill associated virus") OR ("Lymphoid organ virus") OR ("Reo-like virus") OR ("Lymphoid organ vacuolization virus") OR ("Rhabdovirus of penaeid shrimp") OR ("Filamentous bacterial diseas*") OR ("Necrotizing hepatopancreatitis") OR ("Mycobacteriosis") OR ("Chitinolytic bacterial shell diseas*") OR ("Rickettsial infection") OR ("Vibriosis") OR ("Black-gill diseas*") OR ("Hepatopancreatic microsporidiosis") OR ("Black gill syndrome") OR ("Enterocytozoon hepatopenaei") OR ("Cotton shrimp") OR ("Gregarine diseas*") OR ("Black gill diseas*") OR ("Brown gill diseas*") OR ("white spot diseas*") OR ("acute hepatopancreatic necrosis diseas*") OR ("white feces diseas*") OR ("diseas* of shrimp")).

2.2. Analysis Tool

For our assessment of research trends and current status concerning the effects of climate change on global shrimp diseases, we employed CiteSpace. The decision to use CiteSpace stemmed from its adeptness at text-mining, specifically in handling citation metadata and generating visual representations (Noor et al. 2021). Several visualizations emerged from our CiteSpace analysis: Evolution of research outputs, detailing the number of articles, associated authors, contributing institutions, and journals.

It's worth noting that co-citation arises when two distinct paper sources are jointly cited within a single article (Aryadoust and Ang, 2021). Separately, the cluster analysis, sourced from CiteSpace, was utilized to pinpoint prevailing disciplines within our area of focus (Noor et al. 2021).

3. Results

3.1. Trends in Literature

We centered our analysis on scientific publications from the period 1991 to 2022, exploring the nexus between climate change and shrimp diseases (as depicted in Figure 1). Within this timeframe, we pinpointed 400 articles dedicated to shrimp disease research. The annual publication count showed a gradual increment but was interspersed with fluctuations, particularly noticeable between the years 1998 and 2017. A peak in research output was observed in 2020, registering a count of 44 articles.

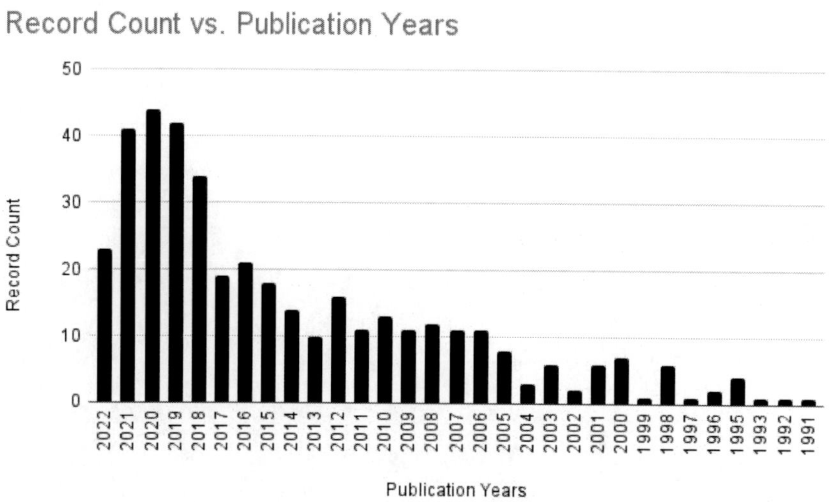

Figure 1. Number of research articles regarding the impacts of climate change on worldwide shrimp diseases.

3.2. Analysis of Countries

Figure 2 illustrates the geographical distribution of articles by countries engaged in research on the interplay between climate change and shrimp diseases. A total of 73 countries contributed to this body of literature. The USA led the pack, accounting for 87 publications (21.75% of the total), trailed by China with 66 articles (16.5%), and Brazil contributing 28 articles, making up 7% of the total publications.

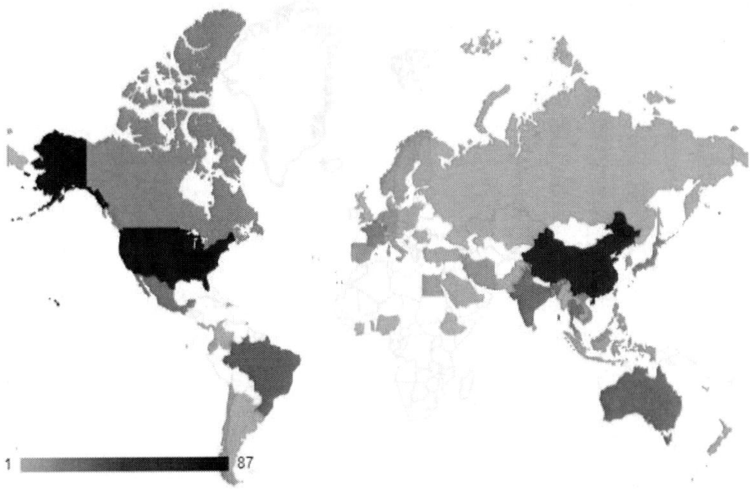

Figure 2. Distribution of national contributions to shrimp diseases research. Dark red signifies countries with the highest volume of publications, while progressively lighter shades indicate decreasing numbers of articles. Nations without any contributions in this specific area are depicted in white.

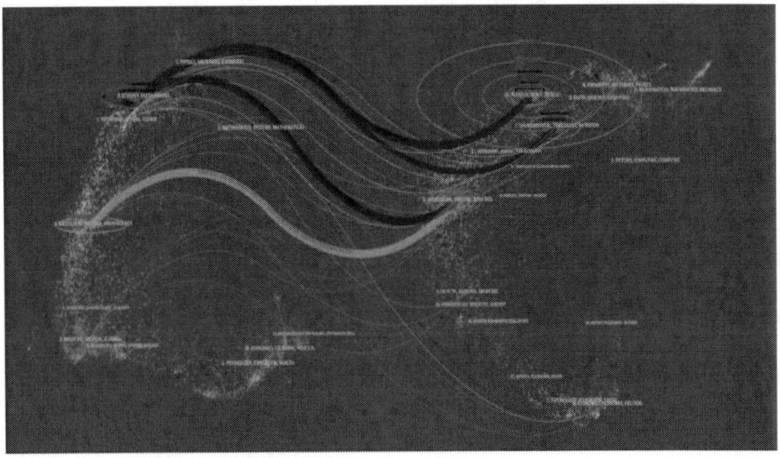

Figure 3. Dual map overlay indicates two different fields (blue and orange) and three different disciplines of research in the areas.

3.3. Dual Map Overlay

Our analysis reveals that within the realm of climate change and shrimp disease, two primary disciplines emerge: "Ecology, Earth, Marine" and "Molecular, Biology, Immunology." Additionally, three associated fields have been identified: "Plant, Ecology, Zoology," "Environmental, Toxicology, Nutrition," and "Molecular, Biology, Genetic" (as illustrated in Figure 3).

3.4. Key Authors in the Field

Marcelo Bahia Labruna, affiliated with the University of São Paulo, Brazil, stands out as the most prolific author in this area, boasting the highest number of publications (as showcased in Figure 4). On the other hand, Donald V. Lightner from the University of Arizona, USA, emerges as the most influential figure in our study (see Figure 5). Cumulatively, the field of climate change and shrimp disease has engaged a diverse pool of 1,899 authors.

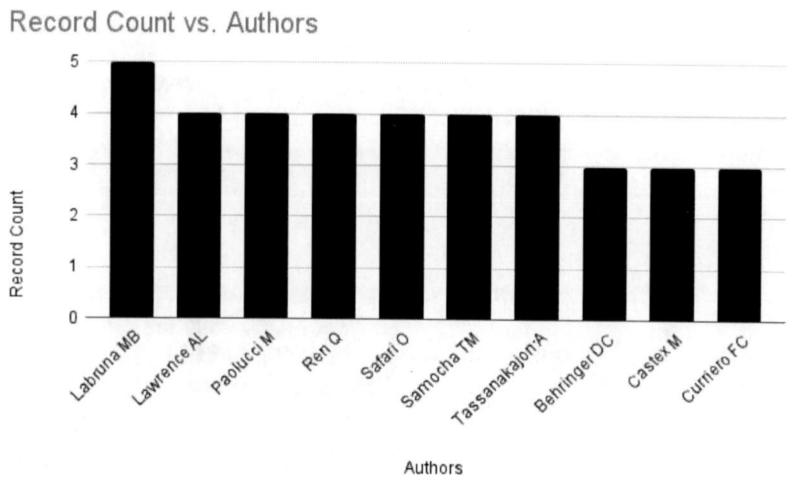

Figure 4. Top 10 authors published in the research of climate change associated with shrimp disease.

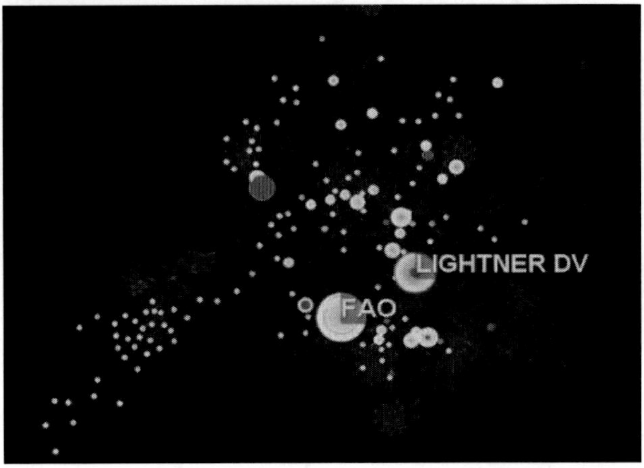

Figure 5. Network of co-cited authors in the field of climate change and shrimp disease.

3.5. Prominent Research Institutions

The Chinese Academy of Sciences, formerly known as Academia Sinica in English (up until 1980), along with the French Research Institute for Exploitation of the Sea (IFREMER), rank as leading institutions worldwide. Both have consistently contributed significant research on the impacts of climate change on shrimp diseases, as illustrated in Figure 6.

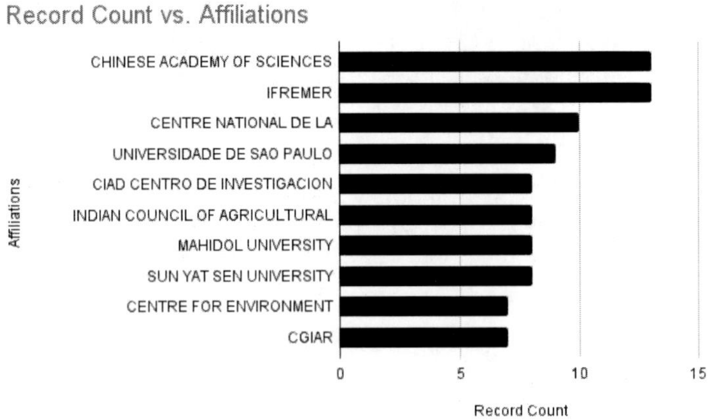

Figure 6. Top ten organizations that had published substantial research on the field of climate change and shrimp disease.

3.6. Leading Journals in the Field

Our findings highlight three predominant journals that have prominently featured research on the intersection of climate change and shrimp disease, as depicted in Figure 7. Interestingly, these very journals also feature prominently in the Journal Citation Analysis network, as derived from the CiteSpace software analysis (see Figure 8).

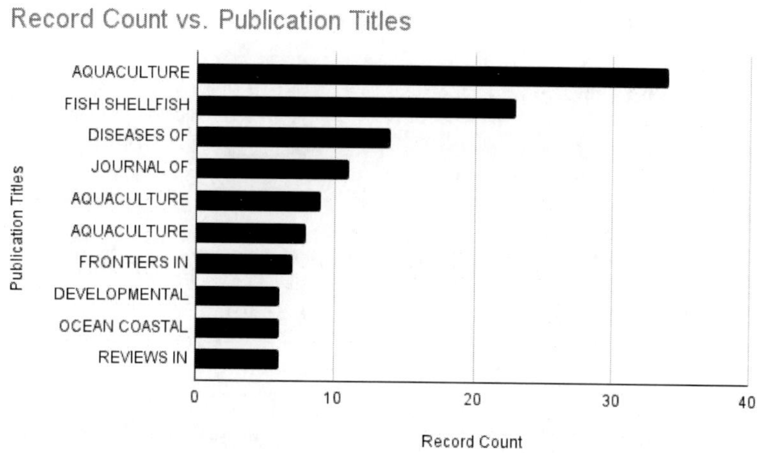

Figure 7. Top ten productive journals based on Web of Science on climate change-related studies towards shrimp disease.

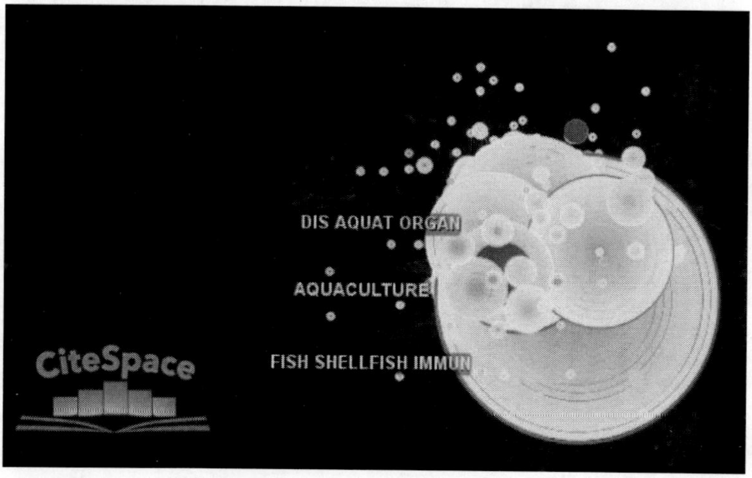

Figure 8. Journal Citation Analysis of Networks.

3.7. Influential Documents and Cluster Analysis

Shields' 2019 study stands out as the sole article fitting the criteria of Document Citation Analysis (DCA) and is recognized as the most influential paper in the domain of climate change and shrimp disease, as showcased in Figure 9. A total of eight distinct clusters have been identified within this field. Of these, #cluster4, labeled "marine parasite," emerges as the newest and exhibits the highest burst. However, the #cluster0, designated "environmental parameter," remains significant, as visualized in Figure 10.

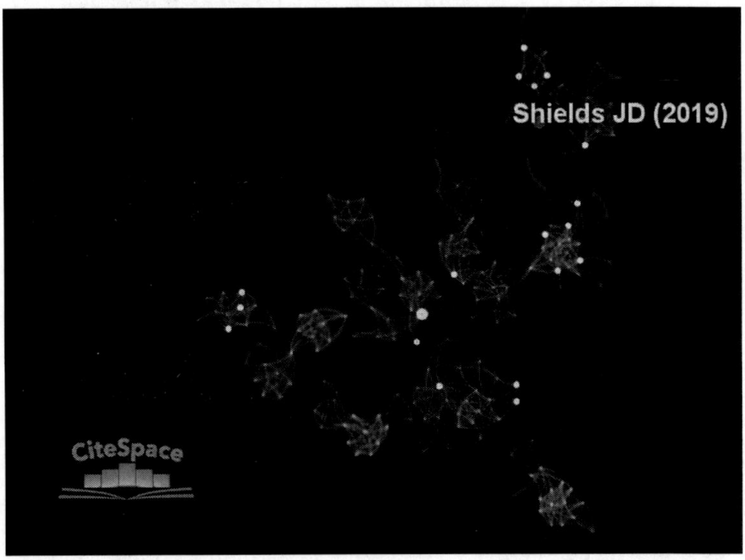

Figure 9. Document Citation Analysis of Networks.

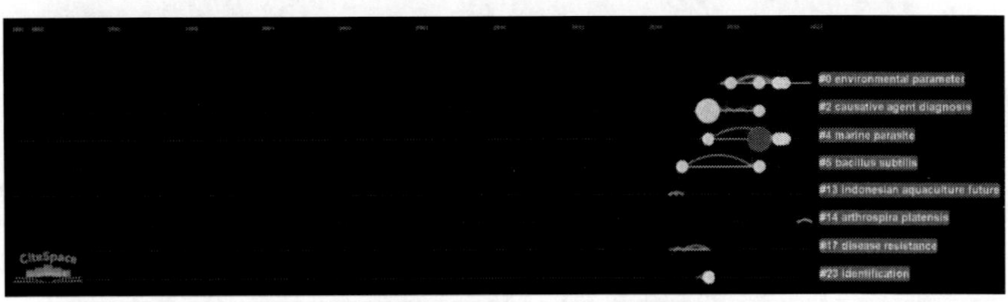

Figure 10. Cluster networks of the research domains, generated from the CiteScpace analysis.

4. Discussion

Climate change has profound impacts on temperature regimes and precipitation patterns throughout the whole world. Global warming accelerates the occurrence of drought and intensify cyclones. Aquaculture sector serves as one of the key anthropogenic activities in contributing to the accumulation of greenhouse gases, including CO_2. The persistent and excessive release of CO_2 has substantially elevated global temperatures. A significant portion of this CO_2 is absorbed by aquatic bodies, causing swift shifts in pH levels. As underscored by Maulu et al. (2021), climate change plays a pivotal role in the sustainability of aquaculture. Extreme pH fluctuations in aquaculture waters induce physiological stress in the cultivated species, compromising their immune defenses and heightening vulnerability to infections. Shrimp farming, akin to other aquaculture practices, faces challenges due to the repercussions of climate change. Moreover, given the soaring demand for shrimp aquaculture as a measure to counter overfishing, the growth of this industry could inadvertently exacerbate climate change effects. In turn, climate change could impact the industry, particularly in terms of revenue, when faced with unforeseen natural calamities such as floods, droughts, or heavy rainfall.

Globally, marine shrimp stands out as the primary cultured species in the shrimp industry. As a hallmark euryhaline species, it possesses a robust osmoregulation adaptive mechanism, allowing it to thrive in diverse salinity ranges—from 1 ppt to 40 ppt—and temperatures between 20-30°C. However, El-Saadony et al. (2022) noted an increased incidence of shrimp diseases when there's a surge in both temperature and salinity. Moreover, a plunge in these parameters due to heavy rainfall can trigger the onset of the White Spot Syndrome Virus (WSSV), leading to a decline in shrimp growth and productivity (Millard et al., 2021).

Extreme pH deviations can impair oxygen affinity, diminish antioxidant capabilities, and amplify disease susceptibility in the shrimp species, *Litopenaeus vannamei*, as highlighted by Chang et al. (2022). When faced with low pH conditions, the environmental bicarbonate essential for shrimp carapace formation dwindles, causing delays in moulting and growth (Chen and Chen, 2003). Therefore, discerning the shrimp's biological reactions to the abiotic shifts induced by climate change is vital. Such insights will inform best aquaculture practices, ensuring a sustainable yield of healthy shrimp. Scientometric analysis emerges as a promising approach to unravel these intricate climate change impacts on shrimp.

As the world grapples with climate change, human and natural environments both need to adjust in response to climatic effects. There is already a wealth of research that attempts to understand how climate change impacts the shrimp aquaculture industry, especially disease-related research. Synthesizing this body of research is important as it provides a critical perspective on current results and insights into gaps for future research. A steady pattern of literature was obtained from the WOSCC database, with 2018 until 2020 indicating an active period for the studies, which also shows the potential of this field or issue in the future. Countries like USA, China and Brazil were among the top countries published in the field. This could be the fact that those countries are among the top producing countries of shrimp in previous decades (Lacerda et al. 2021; FAO, 2020; Biao and Kaijin, 2007). This could lead towards the importance of these data in the future management of shrimp aquaculture-related industries.

Ocean acidification, a significant consequence of global warming, arises from the excessive absorption of atmospheric CO_2 by marine waters. This process disrupts oceanic buffering capacity and alters seawater chemistry, leading to diminished pH levels. Such rapid pH changes can be particularly harmful to penaeid shrimp. Guzmán-Villanueva et al. (2020) noted that these pH variations can intensify disease outbreaks by weakening shrimp's osmotic capacity and immune responses. Furthermore, deviations in pH can impede oxygen affinity, inflict DNA damage, hinder survival, impair antioxidant function, and compromise immunity. These adverse effects culminate in delayed moulting, stunted shrimp growth, heightened vulnerability to pathogens, and even widespread mortalities in *L. vannamei*, as corroborated by both Guzmán-Villanueva et al. (2020) and Li et al. (2019). Given the increasing threats of climate change, including ocean acidification, altered temperature, and changing salinity levels, it's crucial to implement comprehensive management strategies in shrimp aquaculture. This encompasses measures to regulate pH, manage temperature fluctuations, and maintain optimal salinity, ensuring the sustainable and healthy production of shrimp in changing environmental conditions.

Scientometric analyses, which involve quantitative evaluations of scientific literature, offer a comprehensive view of research trends, collaborations, and influential contributors in specific fields. In the context of climate change's impact on shrimp disease, such an analysis reveals a multifaceted research landscape. Chen (2020) and Chen and Leydesdorff (2014) emphasize the value of understanding the interconnections among authors, publications, and journals. This not only provides stakeholders with insights into the evolution of the field over time but also highlights areas of expertise and potential for collaboration.

For instance, the recent work of Noor et al. (2021) offers an overview of the trajectory of aquaculture publications in Southeast Asia. A closer look reveals that leading researchers like Marcelo Bahia Labruna from Brazil and Donald V. Lightner from the USA are making significant contributions, emphasizing the depth of expertise and research capacity of these nations.

Further scrutiny identifies three major journals - "Aquaculture," "Fish and Shellfish Immunology," and "Diseases of Aquatic Organisms" - as primary platforms for disseminating impactful research in this domain. On an institutional front, organizations from China and France, in particular, are leading in terms of publication volume in the WOS database, signaling their active involvement and commitment to advancing the field.

Interestingly, Shields (2019) stands out as a pivotal publication in the WOS database, emphasizing the accelerated disease processes in key crustaceans – such as lobster, crabs, and shrimp – due to climate change. Such findings accentuate the importance of ongoing research and the potential ramifications for the aquaculture industry.

In the research landscape addressing climate change and shrimp disease, cluster analysis has illuminated key focal points. The most cited cluster, #cluster4, centers on the theme of marine parasites. Byers (2020) posits that a significant portion of shrimp diseases can be attributed to parasitic infections, which lends credence to the prominence of this cluster in the current research.

However, when evaluating clusters based on volume—specifically the number of articles listed in the WOS database—#cluster0, themed around environmental parameters, emerges as the predominant focus. Such findings underscore the pivotal role of environmental monitoring in gauging the impact of climate-induced changes on shrimp diseases. As the global climate shifts, partly due to excessive CO_2 emissions, there's a pressing need to understand and respond to its cascading effects on aquatic ecosystems.

Furthermore, the findings accentuate a larger imperative: to enhance the efficiency of global aquaculture practices. By doing so, not only can we mitigate the repercussions of global warming on aquatic life, but also safeguard the sustainability of shrimp production—a critical component in the broader spectrum of global food security.

Conclusion

We conclude that there are two main issues of (i) Increased collaboration between countries to develop global solutions towards shrimp diseases* in the face of climate change and (ii) Drawing from other disciplines outside of "Ecology, Earth, Marine," such as molecular biology, economics, and sustainability, would strengthen and provide additional insights on sustainable shrimp production in the world. Further qualitative discussion was provided along with the future research topic's direction of Sustainable Development Goals of Food Security and Climate Change. Our results suggest that policymakers should focus on local adaptation and empowering decision-making in order to properly adapt to climate change. Our research benefits scientific communities, philanthropic funders, related governments, and NGOs towards sustainably managing ongoing climate change impacts.

Recommandations, Strengths and Limitations

This study offers a comprehensive qualitative discussion, pointing toward the future research directions related to the Sustainable Development Goals on Food Security and Climate Change. To the best of our knowledge, ours is the inaugural research effort to methodically assess and distinguish the data regarding model-based organisms in climate change investigations.

Nonetheless, several limitations persist in the current research. For instance, the software we employed was specifically developed to integrate solely with the WOS database. Furthermore, the English language proficiency presented in the published findings was not without its imperfections, highlighting another limitation.

Aquaculture systems absorb excessive carbon dioxide (CO_2) emissions, which can significantly alter water chemistry and buffering capacities. Such changes can make water conditions unsuitable for shrimp cultivation. The swift variations in water pH, exacerbated by global climate change, impede shrimp immunity and osmoregulation, escalating their vulnerability to diseases. This consistent prevalence of disease might thwart the attainment of optimal yield, subsequently leading to substantial socioeconomic losses for stakeholders in the shrimp industry.

To safeguard the future of the shrimp aquaculture sector, it is crucial to devise strategic business management plans that address the looming challenges, prominently, climate change. The ramifications of climate change on abiotic factors within aquaculture systems are likely to bring about profound socioeconomic repercussions for those dependent on shrimp production. Events such as extreme rainfall can inflict physical damage on shrimp farms, causing shrimp escapes and disruption to both logistics and marketing. Such events can inflict severe financial setbacks on shrimp farmers, with the most severe scenarios pushing them to the brink of bankruptcy. Conversely, rising temperatures and heatwaves pose genuine health threats to shrimp farming communities, manifesting as dehydration and heat strokes among other ailments.

In light of the multifaceted challenges presented by climate change, it becomes imperative to conceptualize a comprehensive adaptation strategy. This involves enhancing and adapting the prevailing shrimp farming technologies and managerial practices.

Impact Statement

Society

Boosting shrimp production augments the profitability of the shrimp industry, enhancing the socioeconomic well-being of farmers and contributing to national food security objectives. Introducing a novel carbon reduction photocatalyst within the shrimp aquaculture system promises not only improved shrimp health but also a decrease in environmental carbon dioxide. This dual effect will have implications for countering climate change, thereby benefiting humans and the broader ecosystem.

Academia

The insights presented across these chapters promise significant additions to existing literature on best practices in aquaculture. More specifically, these findings elucidate potential strategies to preempt and counteract infectious shrimp diseases, like the Early Mortality Syndrome instigated by *Vibrio parahaemolyticus.*

Government

This chapter provides valuable data on methods to prevent shrimp diseases and the interplay of climate change with aquaculture systems. Such insights can guide governmental bodies in crafting robust policies that ensure resilience against climate change within the shrimp farming sector. This, in turn, can catalyze the expansion of the shrimp business, thus enhancing the country's Gross Domestic Product (GDP) and fortifying the national economy. Furthermore, this work aids in comprehending the contemporary dynamics of the shrimp aquaculture landscape, offering policy makers invaluable perspectives that can inform short-term and long-term solutions.

Industry

The shrimp aquaculture sector plays a pivotal role in ensuring food security. Therefore, the formulation of a literary model, underpinned by state-of-the-art scientometric analysis, can provide deeper insights into bolstering shrimp disease resistance, ensuring consistent, high-yield production. Such advancements can further position the national shrimp industry as a dominant player on the global stage.

Funding

The coloured figures for the present chapter has been sponsored by funds provided by USDA-NIFA Sustainable Agriculture Systems, Grant No. 2019-69012-29905 to one of the Editors, Prof. Dr. Guillermo Téllez-Isaías. Title of the project: Empowering US Broiler Production for Transformation and Sustainability USDA-NIFA (Sustainable Agriculture Systems): No. 2019-69012-29905.

Acknowledgments

The study is supported by the Malaysian Ministry of Higher Education under the Long-Term Research Grant Scheme (LRGS/1/2020/UMT/01/1; LRGS UMT Vot No. 56040) entitled 'Ocean climate change: potential risk, impact and adaptation towards marine and coastal ecosystem services in Malaysia' between University Malaysia Terengganu (UMT), University Malaya (UM), University Sains Malaysia (USM) and International Islamic University Malaysia (IIUM). We also acknowledge the funds provided by USDA-NIFA Sustainable Agriculture Systems, Grant No. 2019-69012-29905. Title of Project: Empowering US Broiler Production for Transformation and

Sustainability USDA-NIFA (Sustainable Agriculture Systems): No. 2019-69012-29905.

References

Aryadoust V., Ang B. H. (2021). Exploring the frontiers of eye tracking research in language studies: a novel co-citation scientometric review, *Computer Assisted Language Learning*, 34(7): 898-933. https://doi.org/10.1080/09588221.2019.1647251.

Azra M. N., Mohd Noor M. I., Sung Y. Y., and Abd Ghaffar M. (2022). Visualizing a Field of Research With Scientometrics: Climate Change Associated With Major Aquatic Species Production in the World. F*rontier in Environmental Science*. 10:905428. https://doi.org/10.3389/fenvs.2022.905428.

Azra M. N., Okomoda V. T., Tabatabaei M., Hassan M., and Ikhwanuddin M. (2021). The Contributions of Shellfish Aquaculture to Global Food Security: Assessing Its Characteristics From a Future Food Perspective. *Frontier in Marine Science*. 8:654897. https://doi.org/10.3389/fmars.2021.654897.

Biao X., Kaijin Y. (2007). Shrimp farming in China: Operating characteristics, environmental impact and perspectives, *Ocean & Coastal Management*. 50(7): 538-550. https://doi.org/10.1016/j.ocecoaman.2007.02.006.

Boyd C. E., McNevin A. A., and Davis R. P. (2022). The Contribution of Fisheries and Aquaculture to the Global Protein Supply. *Food Security* 14: 805-827. https://doi.org/10.1007/s12571-021-01246-9.

Braña C. B. C., Kristine C., Senff P., and Stolz I. K. (2021). Towards Environmental Sustainability in Marine Finfish Aquaculture. *Fontiers in Marine Science*. 8: 666662. https://doi.org/10.3389/fmars.2021.666662.

Brooks D. R., Hoberg E. P., Boeger W. A., and Trivellone V. (2022). Emerging Infectious Disease: An Underappreciated Area of Strategic Concern for Food Security. *Transboundary and Emerging Diseases*. 69:254– 267. https://doi.org/10.1111/tbed.14009.

Byers J. E. (2020). Effects of climate change on parasites and disease in estuarine and nearshore environments. *PLoS Biology* 18(11): e3000743. https://doi.org/10.1371/journal.pbio.3000743 .

Chang F., Li N., Shi X., Olga V., Wang X., Diao X., Zhou H., Tang X. (2022). Physiological and muscle tissue responses in Litopenaeus vannamei under hypoxic stress via iTRAQ. *Frontier in Physiology*. 13:979472. https://doi.org/10.3389/fphys.2022.979472.

Chen C., Leydesdorff L. (2014). Patterns of connections and movements in dual-map overlays: A new method of publication portfolio analysis. *Journal of the Association for Information Science and Technology*. 65: 334-351. https://doi.org/10.1002/asi.22968.

Chen C. (2020). *How to Use CiteSpace*. Lean Publishing British Columbia Canada p 246. https://citespace.podia.com/view/downloads/ebook-how-to-use-citespace.

Chen, S. M., Chen, J. C. (2003). Effects of pH on survival, growth, molting and feeding of giant freshwater prawn *Macrobrachium rosenbergii*. *Aquaculture*. 218: 613-623. https://doi.org/10.1016/S0044-8486(02)00265-X.

Dimitrov, R. S. (2016). The Paris agreement on climate change: Behind closed doors. *Global environmental politics*, *16*(3), 1-11.

Doan H. V., Soltani M., Leito A., Shafiei S., Asadi S., Lymbery A. J., and Ringo E. (2022). Streptococcosis a Re-Emerging Disease in Aquaculture: Significance and Phytotherapy. *Animals* 12: 2443. https://doi.org/10.3390/ani12182443.

El-Saadony M. T., Swelum A. A., Abo Ghanima M. M., Shukry M., Omar A. A., Taha A. E., Salem H. M., El-Tahan A. M., El-Tarabily K. A., and El-Hack M. E. (2022). Shrimp Production, the most Important Diseases that Threaten it, and the Role of Probiotics in Confronting these Diseases: A Review. *Reseach in Veterinary Science* 144: 126-140. https://doi.org/10.1016/j.rvsc.2022.01.009.

FAO. (2020). *Regional review on status and trends in aquaculture development in north america*. FAO Fisheries and Aquaculture Circular No. 1232/2. https://www.fao.org/3/cb7659en/cb7659en.pdf.

Flegel, T. W., Lightner, D. V., Lo, C. F., & Owens, L. (2008). Shrimp disease control: past, present and future. *Diseases in Asian Aquaculture VI. Fish Health Section, Asian Fisheries Society, Manila, Philippines*, *505*, 355-378.

Galappaththi, E. K., Ichien, S. T., Hyman, A. A., Aubrac, C. J., & Ford, J. D. (2020). Climate change adaptation in aquaculture. *Reviews in Aquaculture*, *12*(4), 2160-2176.

Guzmán-Villanueva, L. T., Escobedo-Fregoso, C., Barajas-Sandoval, D. R., Gomez-Gil, B., Peña-Rodríguez, A., Martínez-Diaz, S. F., Balcázar, J. L., & Quiroz-Guzmán, E. (2020). Assessment of microbial dynamics and antioxidant enzyme gene expression foll o wing probiotic administration in farmed Pacific white shrimp (*Litopenaeus vannamei*). *Aquaculture,* 519, 734907. https://doi.org/10.1016/j.aquaculture.2019.734907.

Lacerda L. D. D., Ward R. D., Godoy M. D. P., Andrade M. A. J. D., Borges R., and Ferreira A. C. (2021). 20-Years Cumulative Impact From Shrimp Farming on Mangroves of Northeast Brazil. *Frontiers in Forests and Global Change.* 4: 653096. https://doi.org/10.3389/ffgc.2021.653096.

Li, H., Ren, C., Jiang, X., Cheng, C., Ruan, Y., Zhang, X., Huang, W., Chen, T., & Hu, C. (2019). Na+/H+ exchanger (NHE) in Pacific white shrimp (Litopenaeus vannamei): Molecular cloning, transcriptional response to acidity stress, and physiological roles in pH homeostasis. *PLOS ONE,* 14(2), e0212887. https://doi.org/10.1371/journal.pone.0212887.

Maulu S, Hasimuna O. J., Haambiya L. H., Monde C., Musuka C. G., Makorwa T. H. Munganga B. P., Phiri K. J., and Nsekanabo J. D. (2021) Climate Change Effects on Aquaculture Production: Sustainability Implications, Mitigation, and Adaptations. *Frontier in Sustainable Food System.* 5:609097. https://doi.org/10.3389/fsufs.2021.609097.

Millard, R. S., Ellis, R. P., Bateman, K. S., Bickley, L. K., Tyler, C. R., van Aerle, R., Santos, E. M. (2021). How do abiotic environmental conditions influence shrimp susceptibility to disease? A critical analysis focussed on White Spot Disease, *Journal of Invertebrate Pathology*. 186: 107369. https://doi.org/10.1016/j.jip.2020.107369.

Noor M. I. M., Azra M. N., Lim V. C., Zaini A. A., Dali F., Hashim I. M., Hamzah H. C., and Abdullah M. F. (2021). Aquaculture research in Southeast Asia - A scientometric analysis (1990-2019). *International Aquatic Research.* 13: 271-288. https://doi.org/10.22034/IAR.2021.1932503.1166.

Ramli Z., Han D. K., Abdullah F., Rak A. E., and Wei L. S. (2022) The Larval Development of the Asian clam, *Corbicula fluminea* in the Hatchery. *Agriculture reports.* 1(1): 28-38. https://www.multiscipub.com/index.php/AgricultureReports/article/view/21/6.

Rocha C. P., Cabral H. N., Marques J. C., and Gonçalves A. M. M. A. (2022) A Global Overview of Aquaculture Food Production with a Focus on the Activity's Development in Transitional Systems- The Case Study of a South European Country (Portugal). *Journal of Marine Science and Engineering*. https://doi.org/10.3390/jmse10030417.

Rowley A. F., Cross M. E., Culloty S. C., Lynch S. A., Mackenzie C. L., Morgan E., O'Riordan R. M., Robins P. E., Smith A. L., Thrupp T. J., Vogan C. L., Wootton E. C., and Malham S. K. The Potential Impact of Climate Change on the Infectious Diseases of Commercially Improtant Shellfish Populations in the Irish Sea- A Review. *ICES Journal of Marine Science* 71(4): 741-759. https://doi.org/10.1093/icesjms/fst234.

Shields J. D. (2019). Climate change enhances disease processes in crustaceans: case studies in lobsters, crabs, and shrimps. Journal of *Crustacean Biology*, 39 (6): 673–683. https://doi.org/10.1093/jcbiol/ruz072.

Stentiford G. D., Neil D. M., Peeler E. J., Shields E. J., Small H. J., Flegel T. W., Vlak J. M., Jones B., Morado F., Moss S., Lotz J., Bartholomay L., Behringer D. C., Hauton C., and Lightner D. V. (2012). Disease will Limit Future Food Supply from the Global Crustasean Fishery and Aquaculture Sectors. *Journal of Invertebrate Pathology* 110: 141-157. https://doi.org/10.1016/j.jip.2012.03.013.

Chapter 6

Conclusion and Recommendations on Marine Life in Changing Climates

Mohamad Nor Azra[1,2,*]
Hidir Ariffin[3]
and Thirukanthan Chandra Segaran[1]

[1]Institute of Climate Adaptation and Marine Biotechnology (ICAMB), Universiti Malaysia Terengganu (UMT), Terengganu, Malaysia
[2]Research Center for Marine and Land Bioindustry, Earth Sciences and Maritime Organization, National Research and Innovation Agency (BRIN), Pemenang, West Nusa Tenggara, Indonesia
[3]Higher Institution Centre of Excellence (HICoE), Institute of Tropical Aquaculture and Fisheries (AKUATROP), Universiti Malaysia Terengganu, Terengganu, Malaysia

Abstract

Global climate warming exerts a profound influence on myriad oceanic and marine ecosystem processes. This edited collection offers a timely glimpse into contemporary trends and developments concerning the impact of climate change on marine environments and their resources. The series underscores the unique opportunities presented by scientific studies, from selecting pivotal research problems to shaping career paths and advancing within a particular domain. Furthermore, it is anticipated that global warming will directly affect the marine food chain. In summation, this book furnishes a foundational overview of national adaptation policies and practices in the face of climate change.

* Corresponding Author's Email: azramn@umt.edu.my.

In: Marine Life in Changing Climates
Editors: M. Nor Azra, T. Chandra Segaran, G. Téllez-Isaías et al.
ISBN: 979-8-89113-404-1

Keywords: global warming; food web; marine environment; environmental biology

1. Summary of the Book's Key Points and Recommendations

The scientometric analysis presented in this book highlights a notable disparity in the volume of climate change studies across various subjects. Specifically, research on climate adaptation (with approximately 1500 articles in 2020) and molluscs (approximately 2250 articles in 2020) significantly outnumbered those focused on captured fisheries (roughly 375 articles in 2020) and shrimp disease (around 40 journals in 2020). This imbalance underscores the need for heightened scholarly attention to climate change impacts on captured fisheries and especially on shrimp disease. The latter holds particular urgency given the global significance of shrimp as a highly valued crustacean, representing a market value of US$10 billion or 16% of global fishery exports.

Further, the disparity in research volume may spur discussions among researchers about keyword selection strategies. The choice of broader keywords, such as climate adaptation and molluscs, tends to yield higher numbers of relevant articles when scanned by CiteSpace software. Conversely, the use of more specific keywords like captured fisheries and shrimp disease reduces the software's screening capability. However, even though the trend suggests a dominant interest in climate adaptation and molluscs since 2010, it doesn't imply that these areas should monopolize scholarly efforts. There remains a consistent and critical need to pursue research in these domains, as no declining trend in their significance has been observed. Research into climate change reveals a significant concentration of studies stemming from a few key countries: America, China, Australia, Brazil, and Canada. Notably, influential authors cited frequently in the literature are predominantly from these nations. To illustrate, in the field of climate adaptation, leading voices include Hofmann AA, Adger WN, and Azevedo RBR; for captured fisheries, standout contributors are William Cheung, Daniel Pauly, and Xiujuan Chen; within the realm of molluscs, Zhang et al., make a notable impact; and in shrimp disease research, Shields JD is a leading authority.

The prominent research contributions from these countries might be attributed to their economic affluence. Historically, high-income nations have often been in a position to allocate considerable resources to research and development. This investment strategy inherently provides researchers from these countries with greater opportunities to publish and make significant contributions to the scientific literature.

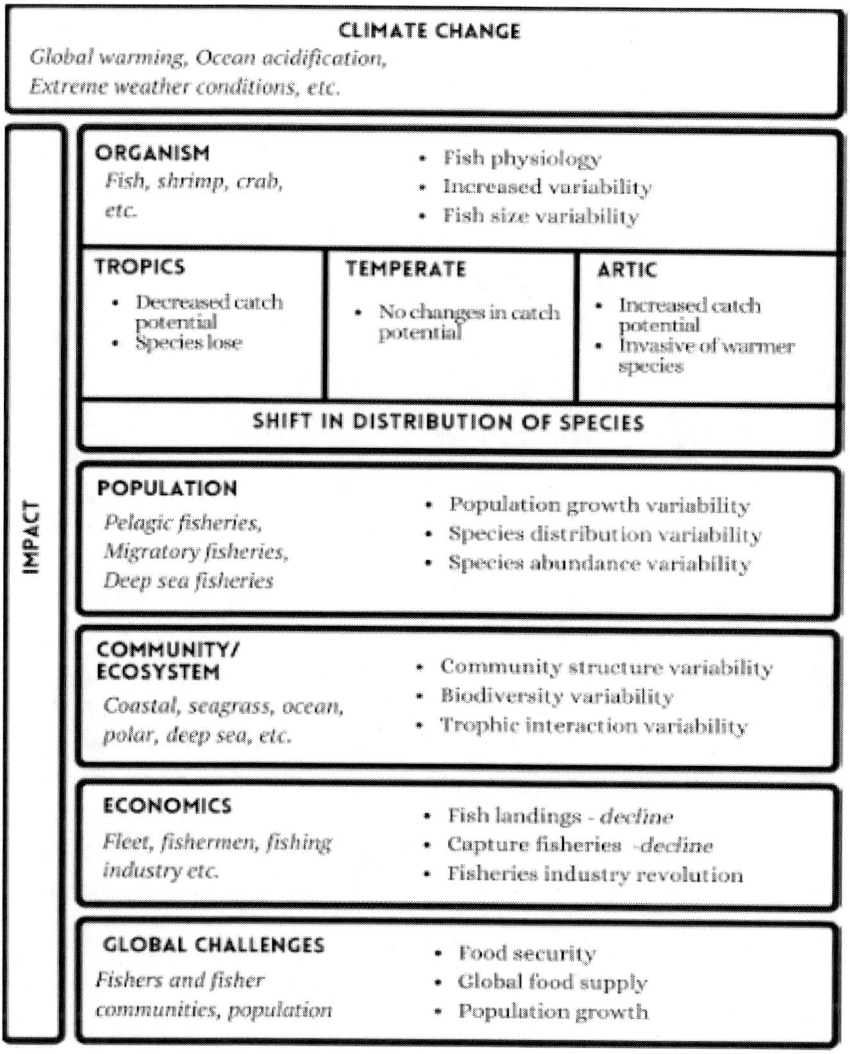

Figure 1. Climate change impacted marine life in various ecosystem and challenges.

Moving forward, it is imperative to establish and encourage more globally inclusive scientific groups. These collectives should aim to foster and promote integrated research endeavors on an international scale. This strategy will not only serve to bolster research in countries with emerging economies but will also work towards achieving a more balanced and comprehensive scientific understanding that spans diverse regions. Such a move is crucial, especially when we consider the global implications of climate change. Without this concerted effort, there's a risk that scientific investigations become too localized, potentially overlooking broader trends and impacts.

In summation, marine ecosystems are undeniably affected by climate change through a variety of mechanisms, as illustrated in Figure 1.

2. Final Thoughts on the Importance of Protecting Marine Life in a Changing Climate

Information compiled in this book is a part of the small effort with the intention to expand the studies pertaining to climate change and marine life. Although this book perhaps merely covered a part of marine life studies, this effort might drive the need for more research to be conducted soon. Through this knowledge, we could build a better reference center for updating any relevant information regarding climate change and continue to predict the far-reaching consequences of climate change on marine life in going forward. And, of course, some future marine life breakthroughs could soon be utilized by the government, stakeholders, and any relevant agency to re-plan and prioritize immediate solutions to resolve the rising climate issues. In winding up, protecting marine life through research efforts reflects our gratitude to deeply appreciate all the luxuries offered by marine ecosystems, which ultimately will end the climate crisis in the near future.

Acknowledgments

The present work was supported by the Long-Term Research Grant Scheme (LRGS) awarded by the Department of Higher Education, Ministry of Higher Education Malaysia (LRGS/1/2020/UMT/01/1; LRGS UMT Vot No. 56040) (2020—2024).

Funding

The edited chapter has been sponsored by USDA-NIFA Sustainable Agriculture Systems, Grant No. 2019-69012-29905 to one of the Editors, Prof.

Dr. Guillermo Téllez-Isaías. Title of the project: Empowering US Broiler Production for Transformation and Sustainability USDA-NIFA (Sustainable Agriculture Systems): No. 2019-69012-29905.

About the Editors

Dr. Mohamad Nor Azra

Institute of Climate Adaptation and Marine Biotechnology (ICAMB),
Universiti Malaysia Terengganu (UMT), 21030,
Kuala Nerus, Terengganu, Malaysia
Email: azramn@umt.edu.my

Dr. Mohamad Nor Azra was born in Port Dickson, Negri Sembilan, in 1987. Dr. Azra received his Diploma, BSc, MSc. and PhD degree in Fisheries and Aquaculture Environment from the Universiti Malaysia Terengganu, Malaysia. He was sponsored by the Malaysian Public Service Department Scholarship, Scholarship Talent Attraction and Retention (STAR), and MyBrain15 (MyPhD & MyMaster) Postgraduate Scholarship during his journey of study from 2011 to 2018. Dr. Azra's current research interest include the adaptation of aquatic animals towards changing climate, focusing on ecophysiological response and behavioural changes. He has over 60 technical publicatios in these areas, including 30 international refereed journal papers, many of which are well cited. He is currently the inaugural fellow for Leadership for Climate Resilient Fisheries Fellowship, USA and Visiting Researcher at National Research and Innovation Agency, Indonesia.

Dr. Thirukanthan Chandra Segaran

Institute of Climate Adaptation and Marine Biotechnology (ICAMB),
Universiti Malaysia Terengganu (UMT), 21030,
Kuala Nerus, Terengganu, Malaysia
Email: thiru@umt.edu.my

Dr. Thirukanthan serves as the Chief Senior Science Officer at Universiti Malaysia Terengganu's prestigious Institute of Climate Adaptation and Marine Biotechnology, where he spearheads the Core Facilities and directs the institute's laboratory and technical team. A marine biology graduate, he further elevated his academic credentials with a Master of Science and a PhD in Biotechnology. His prowess earned him the distinguished ASEM-Duo research fellowship, facilitating cutting-edge research on microalgae biofilm development at the University of South Bretagne, France. Dr. Thirukanthan has been the recipient of eleven international and seventeen national awards for his inventions from various research exhibitions. His standout invention, Scalogel, aimed at advanced wound healing, not only clinched the World International Intellectual Property Organization Award at Geneva's 39th International Exhibition of Inventions but also secured a gold medal and the esteemed IENA Germany Special Award in the realm of international invention and innovation.

Professor Dr. Guillermo Téllez-Isaías

Department of Poultry Science, University of Arkansas,
Fayetteville, AR 72701, USA
Email: gtellez@uark.edu

Guillermo Tellez-Isaias was born in Mexico City, in 1963. He received his Doctor in Veterinary Medicine degree in 1986 and his Master in Science degree in Veterinary Sciences in 1989 from the National Autonomous University of Mexico (UNAM), and his PhD from Texas A & M University. He worked as full Professor at UNAM for 16 years, 8 as head of the Avian Medicine Department at the College of Veterinary Medicine. Tellez was President of the National Poultry Science Association or Mexico during 1997-1998; is member of the Mexican Veterinary Academy and the Mexican National Research System. In 2001, Tellez worked during his sabbatical as Visiting Scientist at the Center of Excellence in Poultry Science of the University of Arkansas, where is still working as Research Professor. His research is focused on the advantages of the poultry gastrointestinal model to evaluate the beneficial effects of functional foods to enhance intestinal health and disease resistance.

Dr. Hidir Ariffin

Higher Institution Centre of Excellence (HICoE), Institute of Tropical Aquaculture and Fisheries, Universiti Malaysia Terengganu, 21030, Kuala Nerus, Terengganu, Malaysia
Email: hidirariffin@umt.edu.my

Dr. Hidir Ariffin was born in Klang, Selangor, Malaysia, in April 1991. Later, he migrated to Terengganu, Malaysia, which in this homeland he highly expressed his thankful to Universiti Malaysia Terengganu for this beloved place where he opted and executed his higher education journey from a Bachelor's degree (Marine Biology), Master's degree (Aquaculture) until achieving Doctor of Philosophy (Aquaculture). Currently, he works as a Postdoctoral at the Institute of Tropical Aquaculture and Fisheries (AKUATROP) at Universiti Malaysia Terengganu, and is highly keen on any research related to the crustacean field. Going forward, he hoped that he could disseminate knowledge on the crustacean field and resolve any rising issues regarding crustaceans in tandem with fostering humans to treasure our biodiversity germs, especially crustacean life.

Index